AI and Digital Technology for Oil and Gas Fields

This book essentially covers the growing role of AI in the oil and gas industry, including digital technologies used in the exploration phase, customer sales service, and cloud-based digital storage of reservoir simulation data for modeling. It starts with the description of AI systems and their roles within the oil and gas industry, including the agent-based system, the impact of industrial IoT on business models, and the ethics of robotics in AI implementation. It discusses incorporating AI into operations, leading to the reduction of operating costs by localizing control functions, remote monitoring, and supervision.

Features of this book are given as follows:

- It is an exclusive title on the application of AI and digital technology in the oil and gas industry.
- It explains cloud data management in reservoir simulation.
- It discusses intelligent oil and gas well completion in detail.
- It covers marketing aspects of oil and gas business during the exploration phase.
- It reviews development of digital systems for business purposes.

This book is aimed at professionals in petroleum and chemical engineering, technology, and engineering management.

AI and Digital Technology for Oil and Gas Fields

Niladri Kumar Mitra

CRC Press is an imprint of the
Taylor & Francis Group, an **informa** business

Designed cover image: shutterstock

First edition published 2025
by CRC Press
2385 NW Executive Center Drive, Suite 320, Boca Raton FL 33431

and by CRC Press
4 Park Square, Milton Park, Abingdon, Oxon, OX14 4RN

CRC Press is an imprint of Taylor & Francis Group, LLC

ISBN: 9781032309897 (hbk)
ISBN: 9781032310329 (pbk)
ISBN: 9781003307723 (ebk)

DOI: 10.1201/9781003307723

Typeset in Times
by codeMantra

Contents

About the Author

Niladri Kumar Mitra graduated in Petroleum Engineering from IIT-Indian School of Mines, Dhanbad, in 1972. He is a performance-driven, self-motivated professional with proven leadership, administrative, managerial, and technical skills of over 50 years with ONGC, a major Indian national oil and gas company, Reliance Industries Limited, the largest private sector E&P company, Gujarat State Petroleum Corporation (GSPC), and McDermott. He is vastly acknowledged and credited for putting forward his acquired experience, expertise, wisdom, world view, emerging paradigms, and perspectives in its right earnest to ensure the best energy security for India and worldwide. He was the Director (Offshore) of ONGC Ltd., India, during the initial phase of Mumbai High Development. He is among the few frontrunners in interactive knowledge sharing and mentoring. He has chaired and steered many international and national conferences and has put forward many new technologies and paradigms for oil and gas industries through his technical articles, keynote addresses, and presentations. He introduced new cutting-edge technologies in the industry such as digital process control, intelligent well completion, intelligent platform, and hydrofracturing of high pressure and high temperature (HPHT) well. His first floating production storage and offloading (FPSO) at D1 was conceptualized, designed, and awarded.

He was the Chief Operating Officer, GSPC E&P Group, looking after the Development of Deen Dayal Field (DD Field) in the KG offshore basin and he was overall accountable for the oil and gas field, one of the HPHT natural gas Deen Dayal Fields of the KG Basin. He led from the front in designing, construction, commissioning, and operation including designing of HT/HP wells for the first time in India.

He served as President (E&P) of Accountabilities at Reliance Industries Limited. His role lies in giving the organization an edge at the offshore front by managing the roles and responsibilities of a complete vertical hierarchy of deepwater offshore assets, including looking after the development, drilling, and completion of D-6, MA field including sub-sea completion.

He was the Chairman of the Society of Petroleum Engineers for 7 years. He has authored *Fundamentals of Floating Production System, Principle of Artificial Lift, One Mile in Oil and Gas*, and *Hydro Fracturing of HPHT Oil and Gas Well*, which are acclaimed by the oil and gas industry and petroleum engineering institutes. He worked as an adjunct professor in IIT for 10 years.

Preface

This book tries to introduce artificial intelligence (AI) as the ability of a computer or a robot controlled by a computer to perform tasks that humans usually do because they require human intelligence and discernment. The primary objective of AI (also called heuristic programming, machine intelligence, or the simulation of cognitive behavior) is to enable computers to perform such intellectual tasks as decision-making, problem-solving, perception, and understanding human communication (in any language, and translating them). AI is a theory and development of computer systems that can perform tasks that normally require human intelligence. Speech recognition, decision-making, and visual perception, for example, are features of human intelligence that AI may possess. AI is also dipping into the digital oilfield well to leverage the latest advancement in intelligent visual monitoring technology to give companies the ability to limit remote site visits while reducing operational expenditure and mitigating safety and environmental risks. Many analog signals—temperature, pressure, strain, and so on—can benefit from some conditioning to improve measurement quality. Filtering is a type of conditioning that removes interference from system signals—also considered knowledge base system and data filtering.

This book also discusses how, in terms of interaction, contribution, and dependency, agent-based systems and AI are closely related. An agent such as a human or a robot identifies the environment through sensors and effectors. It uses a search and pattern matching method where the computer is instructed to search its knowledge base on the match found and if specific conditions are met to solve a problem. A user interface that involves some aspects of AI (or computational intelligence) is referred to as an intelligent user interface (IUI). A user model and deep domain knowledge are typically required for an IUI on the computer side.

Automation has introduced a system of computers and machines and replaced it with a system built by combining man and machine. Highly intense and repetitive tasks have become efficient, and the product quality has also increased with automation in various industries. Before diving into AI in automation, it's essential to acknowledge that both these terms are used interchangeably in daily life.

The components of trusted AI simulate human intelligence processes by machines, especially computer-based systems. These processes generally include learning, reasoning, and self-correction. Some of the applications of AI include expert systems and machine vision, bringing trust back into AI through open source. Increasingly, dimensions of faith, including fairness, robustness, and explainability, are important metrics that help to evaluate AI model behavior. AI systems are increasingly being used to support human decision-making. While AI holds the promise of delivering valuable insights and knowledge across many applications, the broad adoption of AI systems will rely heavily on the ability to trust their output.

The Internet of Things (IoT) describes the extension of internet connectivity between physical devices and everyday objects. Embedded with electronics, internet connectivity, and other forms of hardware such as sensors, these devices can communicate and interact over the internet and can be remotely controlled and monitored. The Industrial Internet of Things (IIoT) impacts established manufacturing companies' business models (BM) within several industries. Thus, it is being tried to analyze the influence of the IIoT on these BMs, particularly with respect to differences and similarities dependent on varying industry sectors.

Robotic process automation (RPA) is a game-changing technology designed to automate high-volume, repeatable tasks that take up a large percentage of a worker's time. Industrial robots have automated blue-collar work in factories. RPA software is intended to automate mostly office work. There are five significant fields of robotics, namely, human–robot interface, mobility, manipulation, programming, and sensors, and they are important for robotics development.

Machine learning is part of AI. It completes learning from data with specific inputs to the machine. Without explicit programming, computers can learn how to complete tasks using machine learning. Machine learning involves computers learning from data provided to carry out specific tasks. It is an application of AI that involves algorithms and data that automatically analyze and make decisions by itself without human intervention. It describes how a computer performs tasks on its own using previous experiences. User feedback should surround the generalization of the machine learning process. Machine learning contains algorithms for supervised learning, unsupervised learning, and semi-supervised learning, which set them apart from one another. Machine learning is used for speech recognition, and acoustic models are used for fraud detection in credit cards. Three methods, i.e., statistical methods, computer science, and economics-biology-psychology, are brought together by machine learning. This book will not deal with statistical methods.

Intelligent drilling and well completion is a relatively independent work linking drilling and production engineering, combining engineering, geology, and exploration. Many major oil and gas fields have gradually entered into the deepwater development stage, and many unconventional reservoirs such as shale gas also launched large-scale development. Intelligent well completion is a complete system of the production well that enables continuous and real-time reservoir management. The core of the technology is to form a closed-loop control; therefore, data such as downhole temperature, pressure, flow, and the composition of fluids collected by the wellbore sensor are fed back to the uphold system in real time.

A 3D seismic survey impacts the original oil and gas field development plan. With the drilling of new development wells, the added information is used to refine the actual interpretation of the subsurface data collected. With the passage of time data is built, elements of the 3D data that are initially ambiguous make some sense. The usefulness of a 3D seismic survey lasts for the life of the oil and gas reservoir. The most popular technology is aiding the transformation toward automation and efficiency in cloud computing for storing massive data, which is frequently used for reservoir simulation of oil and gas reservoirs. In the cloud platform system, high-performance computing (HPC) is meant to provide the capability of scaling to large numbers of tasks that run in parallel on demand. However, it remains a dilemma for companies in the oil and gas industry whether to transition HPC activities to the cloud or not. DEFI software is used to optimize field, production, and reservoir data storage in iCloud storage.

A distributed control system (DCS) is a computerized control system for an oil and gas process plant and drilling rig, usually with many control operational loops. Autonomous controllers are distributed throughout the system. This contrasts with systems that use centralized controllers; discrete controllers are located either at a central control room or within the main computer. DCS can also monitor and control through human–machine interface (HMI), which provides sufficient data to the operator to take charge over various processes, and it acts as the heart of the system. The control is distributed throughout using a powerful and secure communication system. The various process inputs are connected to the controllers directly or through IO bus systems such as Profibus and Foundation Fieldbus.

Significant progress has been made through various initiatives taken by the Indian government to develop technologies in the oil and gas exploration and production sectors. Different types of new technological developments have come up in seismic shooting and data collection, modern oil and gas metering, hydrofracturing of oil and gas horizons, and software modeling. A massive transformation to a new type of business called e-business containing e-signature, e-invoice, e-commerce, internet, mobile banking, and e-payments creates efficiency in corporate and individual life. Digital platforms will become the new business model and accelerate the reach of new markets. AI will have a significant influence on current technologies. This book will deal with oil and gas exploration, refinery operational optimization, and maintenance with the software application, product distribution, and e-invoicing.

I am happy to receive the suggestions and critical reviews of this book. This book will be very useful for beginners, especially for oil and gas industry personnel and students.

Niladri Kumar Mitra
January 17, 2022

Acknowledgments

I am pleased to present my fifth book after four consecutive best sellers in Petroleum Engineering technical publications. This book gives readers firsthand experience regarding the application of artificial intelligence (AI) in our day-to-day life with the increasing awareness of the Internet of Things. The book's flow is simple, the concepts are illustrative, and the coverage is quite comprehensive. It provides an implicit understanding of the various facets of AI implementation through a case study that shows how a major Latin American online retailer headquartered in Buenos Aires had improved the company's productivity and efficiency after incorporating AI in their business. Although several books and journals are available on AI, this is a single comprehensive book combining intelligent drilling and completion, along with the application of digital control, which has been written in plain language in order to be understandable by students in academics who have not seen it in their fields and do not have hands-on experience. This book also deals with machine learning and robotics' role in AI.

I thank A. Basu for his special support in reviewing the chapter on digital control system (Chapter 10), which has scaled a novel level in superiority. During the writing of this book, I had very close interactions with my industry peers, who provided me with innovative information on various topics in this book.

I wrote the book over 2 years. I spent much of that time in closed doors of my house during the pandemic, and since then, it has been sleepless nights for the rest of the family. I am fortunate to have such a supportive family who always encourage me to pursue the dedicated author within me.

Niladri Mitra
Date: January 24, 2022
Place: Mumbai

1 The Growing Role of Artificial Intelligence in the Oil and Gas Industry

1.1 HISTORY AND TRENDS OF FUTURE DIGITAL TECHNOLOGY

Artificial intelligence (AI) is a commonly employed appellation to refer to the field of science aimed at providing machines with the capacity of performing functions such as logic, reasoning, planning, learning, and perception. The term 'artificial intelligence' was closely associated with the field of "symbolic AI," which was popular until the end of the 1980s. To overcome some of the limitations of symbolic AI, sub-symbolic methodologies such as neural networks, fuzzy systems, evolutionary computation, and other computational models started gaining popularity, leading to "computational intelligence," a subfield of AI. As the hype around AI has accelerated, vendors have been scrambling to promote how their products and services use AI. Often, what they refer to as AI is simply one component of AI, such as machine learning. AI requires a foundation of specialized hardware and software for writing and training machine learning algorithms. No one programming language is synonymous with AI, but a few, including Python, R, and Java, are popular.

In general, AI systems work by ingesting large amounts of labeled training data, analyzing the data for correlations and patterns, and using these patterns to make predictions about future states. In this way, a chatbot that is fed examples of text chats can learn to produce lifelike exchanges with people, or an image recognition tool can learn to identify and describe objects in images by reviewing millions of examples.

Nowadays, AI encompasses the whole conceptualization of a machine that is intelligent in terms of both operational and social consequences. A practical definition used is the one proposed by Mr. Russell and Mr. Norvig: "Artificial intelligence is the simulation of human intelligence processes by machines, especially computer systems." Specific applications of AI include expert systems, natural language processing, speech recognition, and machine vision, which will be discussed here.

Compared to the human intellect, AI is much better at predicting behavior. AI has the capacity to create value systems that are incomprehensible, leading to questionable decisions that seriously impact human lives. That's scary. But there's a way to control that—it's called a transparency switch, which includes the following features:

DOI: 10.1201/9781003307723-1

- Avoiding regulatory and ethical issues
- Optimizing AI through control and collaboration
- Creating a robust AI quality assurance process
- Harnessing the power of AI to deliver real business value

Current AI technologies are used in online advertising, driving, aviation, medicine, and personal assistance image recognition. The recent success of AI has captured the imagination of both the scientific community and the public. An example of this is vehicles equipped with an automatic steering system, also known as autonomous cars. Each vehicle is equipped with a series of LIDAR (light detection and ranging) sensors and cameras, which enable the recognition of its three-dimensional environment and provide the ability to make intelligent decisions on maneuvers in variable, real-traffic road conditions. Another example is the AlphaGo, developed by Google Deep Mind to play the board game Go. Recently, Alpha-Go defeated the Korean grand master Mr. Lee Sedol, becoming the first machine to beat a professional player, and recently, it went on to win against the recent world number one, Mr. Ke Jie, in China. The number of possible games in Go is estimated to be 10,761, and given the extreme complexity of the game, most AI researchers believed it would be years before this could happen. However, current AI technologies are limited to very specific applications. One limitation of AI is the lack of common sense, i.e., the ability to judge information beyond its acquired knowledge. A recent example is that of the AI robot Tay, designed and developed by Microsoft for making conversations on social networks. It had to be disconnected shortly after its launch because it was unable to distinguish between positive and negative human interactions. AI is also limited in terms of emotional intelligence. AI can only detect basic human emotional states such as anger, joy, sadness, fear, pain, stress, and neutrality. Emotional intelligence is one of the next frontiers of higher levels of chronological development of AI-structured Chart personalization (Figure 1.1). At this level, AI will mimic human cognition to a point that it will enable the ability to dream, think, feel emotions, and have own goals. AI research has tried and discarded many different approaches during its lifetime, including simulating the brain, modeling human problem-solving, formal logic, large databases of knowledge, and imitating animal behavior. In the first part of the 21st century, highly mathematical statistical machine learning has dominated the field, and this technique has proved to be highly successful in solving many challenging problems in both industry and academia.

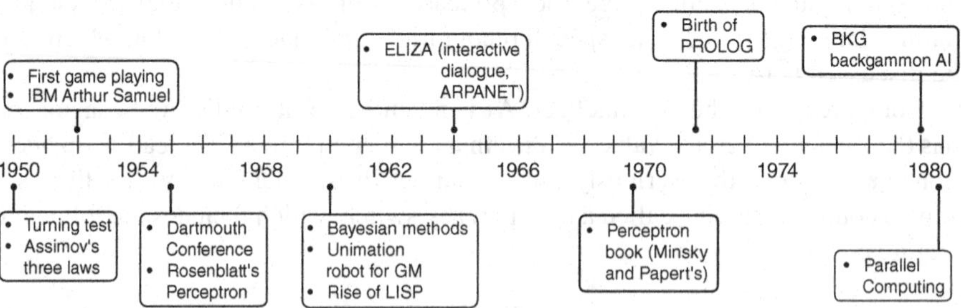

FIGURE 1.1 Chronological development of AI.

1.2 AI TECHNIQUES

In the real world, the following are some unwelcomed properties of AI:

- Its volume is huge, next to unimaginable.
- It is not well organized or well formatted.
- It keeps changing constantly.

AI technique is a manner to organize and use knowledge efficiently in such a way that:

- It should be perceivable by the people who provide it.
- It should be easily modifiable to correct errors.
- It should be useful in many situations, though it is incomplete or inaccurate.

AI techniques elevate the speed of execution of the complex program it is equipped with.

1.2.1 Applications of AI

AI is relevant to any intellectual task. Modern AI techniques are pervasive and are too numerous to list here. Frequently, when a technique reaches mainstream use, it is no longer considered AI; this phenomenon is described as the AI effect.

With social media sites overtaking TV as a source of news for young people and news organizations increasingly reliant on social media platforms for generating distribution, major publishers now use AI technology to post stories more effectively and generate higher volumes of traffic.

AI can also produce deep fakes, a content-altering technology. ZDNet reports, "It presents something that did not actually occur." Although 88% of Americans believe deep fakes can cause more harm than good, only 47% of them believe they can be targeted.

AI has been dominant in various fields (machine learning process) given as follows:

- **Gaming**—AI plays a crucial role in strategic games such as chess, poker, and tic-tac-toe, where a machine can think of a large number of possible positions based on heuristic knowledge.
- **Natural Language Processing**—It is possible to interact with the computer that understands natural language spoken by humans.
- **Expert Systems**—There are some applications that integrate machine, software, and special information to impart reasoning and advising. They provide explanation and advice to the users.
- **Vision Systems**—These systems understand, interpret, and comprehend visual input on the computer. For example:
  - A spying airplane takes photographs, which are used to figure out spatial information or map of the areas.

- Doctors use clinical expert system to diagnose the patient.
- Police department uses computer software that can recognize the face of criminal with the stored portrait made by forensic artist.

- **Speech Recognition**—Some intelligent systems are capable of hearing and comprehending the language in terms of sentences and their meanings while a human talks to it. It can handle different accents, slang words, noise in the background, change in human's noise due to cold, etc.
- **Handwriting Recognition**—The handwriting recognition software reads the text written on paper with a pen or on screen with a stylus. It can recognize shapes of letters and convert them into editable text.
- **Intelligent Robots**—Robots are able to perform the tasks given by a human. They have sensors to detect physical data from the real world such as light, heat, temperature, movement, sound, bump, and pressure. They have efficient processors, multiple sensors, and huge memory, to exhibit intelligence. In addition, they are capable of learning from their mistakes, and they can adapt to the new environment.

AI is important because it can give enterprises insights into their operations that they may not have been aware of previously, and because, in some cases, AI can perform some tasks better than humans. Particularly when it comes to repetitive, detail-oriented tasks like analyzing large numbers of legal documents to ensure relevant fields are filled in properly, AI tools often complete jobs quickly, with relatively few errors. This has helped fuel an explosion in efficiency and opened the door to entirely new business opportunities for some larger enterprises. Prior to the current wave of AI, it would have been hard to imagine using computer software to connect riders to taxis, but today Uber has become one of the largest companies in the world by doing just that. It utilizes sophisticated machine learning algorithms to predict when people are likely to need rides in certain areas, which helps proactively get drivers on the road before they're needed. As another example, Google has become one of the largest players in providing a range of online services by using machine learning to understand how people use their services and then improving them.

1.2.1.1 Advantages and Disadvantages of AI

Artificial neural networks and deep learning AI technologies are quickly evolving, primarily because AI processes large amounts of data much faster and makes predictions more accurately than humanly possible. While the huge volume of data being created on a daily basis would bury a human researcher, AI applications that use machine learning can take that data and quickly turn it into actionable information. As of this writing, the primary disadvantage of using AI is that it is expensive to process the large amounts of data that AI programming requires.

Advantages:

- Good at detail-oriented jobs
- Reduced time for data-heavy tasks
- Consistent results delivery
- Availability of AI-powered virtual agents

Disadvantages:

- Expensive
- Deep technical expertise required
- Limited supply of qualified expert to build AI tools
- Only knows what it's been shown
- Lack of ability to generalize from one task to another

Strong AI vs. Weak AI:

AI can be categorized as either weak or strong.

Weak AI, also known as narrow AI, is an AI system that is designed and trained to complete a specific task. Industrial robots and virtual personal assistants, such as Apple's Siri, use weak AI. Strong AI, also known as artificial general intelligence, describes programming that can replicate the cognitive abilities of the human brain. When presented with an unfamiliar task, a strong AI system can use fuzzy logic to apply knowledge from one domain to another and find a solution autonomously. In theory, a strong AI program should be able to pass both a Turing test and the Chinese room test.

1.2.2 Types of AI Systems

Mr. Arend Hintze, an assistant professor of integrative biology and computer science and engineering at Michigan State University, explained in a 2016 article that AI can be categorized into four types, beginning with the task-specific intelligent systems in wide use today and progressing to sentient systems, which do not yet exist. The various categories of AI are as follows (Table 1.1):

- **Type 1: Reactive Machines.** These AI systems have no memory and are task-specific. An example is Deep Blue, the IBM chess program that beat Mr. Garry Kasparov in the 1990s. Deep Blue can identify pieces on the chessboard and make predictions. Since it has no memory, it cannot use past experiences to inform future ones.
- **Type 2: Limited Memory.** These AI systems have memory, so they can use past experiences to inform future decisions. Some of the decision-making functions in self-driving cars are designed this way.
- **Type 3: Theory of Mind.** Theory of mind is a psychology term. When applied to AI, it means that the system would have the social intelligence to understand emotions. This type of AI will be able to infer human intentions and predict behavior, a necessary skill for AI systems to become integral members of human teams.
- **Type 4: Self-Awareness.** In this category, AI systems have a sense of self, which gives them consciousness. Machines with self-awareness understand their own current state. This type of AI does not yet exist.

In R&D sector, AI had focused chiefly on the following components of intelligence: learning, reasoning, problem-solving, perception, and language understanding.

TABLE 1.1
Different Role of AI Systems

Reactive AI	Limited Memory	Theory of Mind	Self-Aware
Good for simple classification and pattern recognitions	Can handle complex classification tasks	Able to understand human motives and reasoning. Can deliver personal experience to everyone based on their motives and needs	Human-level intelligence that can bypass our intelligence too
Great for scenario where all parameters are known: it can make calculations much faster	Able to make prediction from historical data	Able to learn with fewer examples because it understands motive and intent	
Incapable of dealing with scenarios including imperfect information or requiring historical understanding	Capable of complex task such as self-driving cars, but still vulnerable to outliners or adversarial example	Considered the next Pillars-structure for AI's evaluation	
	Some claim that we have reached a wall with AI in its current condition		

1.3 HARDWARE FOR AI

In 1965, Mr. Gordon Moore observed that the number of transistors, in a dense integrated complex circuit, doubles approximately every year. In 2015, Mr. Moore realized that the rate of progress in the hardware would reach saturation and the transistors would arrive at the limits of miniaturization at the atomic level. Experts estimate that Moore's law could end in 2025. Today, his prediction is still valid and the number of transistors is increasing even if, after 2005, the frequency and the power started to reduce, leading to a core scaling rather than a frequency improvement. Therefore, since 2005, the market is no longer manufacturing faster computers, but the hardware is designed in a multicore manner. To take full advantage of the different types of hardware implementation, the software has been written in a multi-threaded manner. In future, experts believe that revolutionary technologies may help sustain Mr Moore's law. One of the key challenges will be the design of gates in nanoscale transistors and the ability of controlling the current flow as, when the device dimension shrinks, the connection between transistors becomes more difficult and complicated.

Modern computers combine powerful multicore CPUs with dedicated hardware designed to solve parallel processing. Graphics Processing Unit (GPU) and Field Programmable Gate Array (FPGA) are the most popular dedicated hardware commonly available in workstations developing AI systems. A GPU is a chip designed to accelerate the processing of multidimensional data such as an image. A GPU consists of thousands of smaller cores, intended to work independently on a subspace of the input data, that need heavy computation. Repetitive functions that can be applied to different parts of the input, such as texture mapping, image rotation, translation, and filtering, are performed at a much faster rate and more efficiently, through the use of the GPU. A GPU has dedicated memory, and the data must be moved in and out in order to be processed.

FPGA is a reconfigurable digital logic containing an array of programmable logic blocks and a hierarchy of reconfigurable interconnections (Figure 1.2). It is an integrated circuit that can be programmed by a user for a specific use after it has been manufactured. It is not a processor, and therefore, it cannot run a program stored in the memory. FPGA chips enable expert to reprogram logic gates. Expert can use FPGA technology to overwrite chip configurations and create customary circuits. FPGA chips are especially useful for machine learning and deep learning. An FPGA is configured using a hardware description language (HDL), and unlike the traditional CPU, it is truly parallel. This means that each independent processing task is assigned to a dedicated section of the chip, and many parts of the same program can be performed in simultaneously.

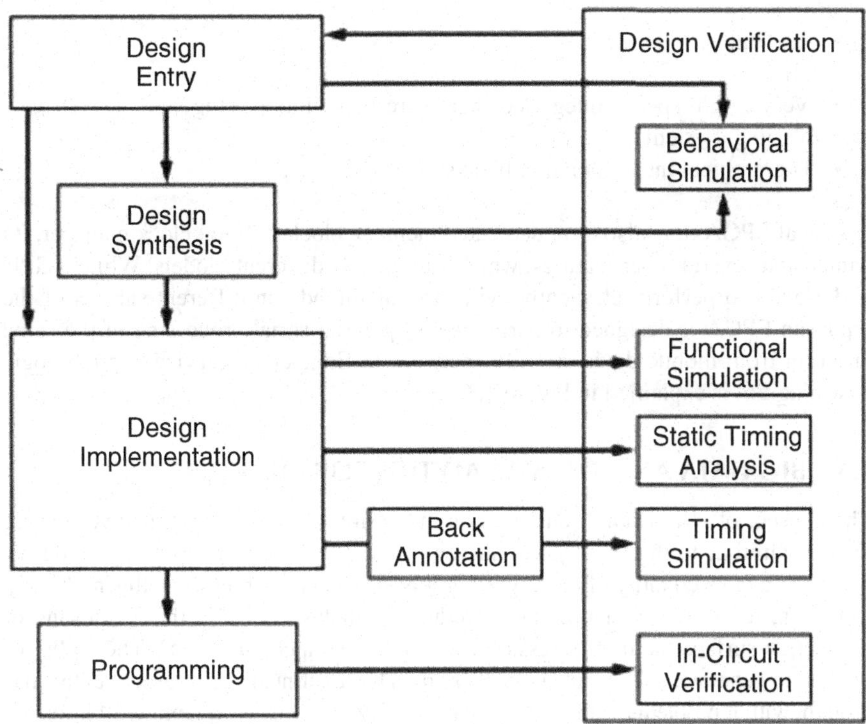

FIGURE 1.2 FPGA programmable logic.

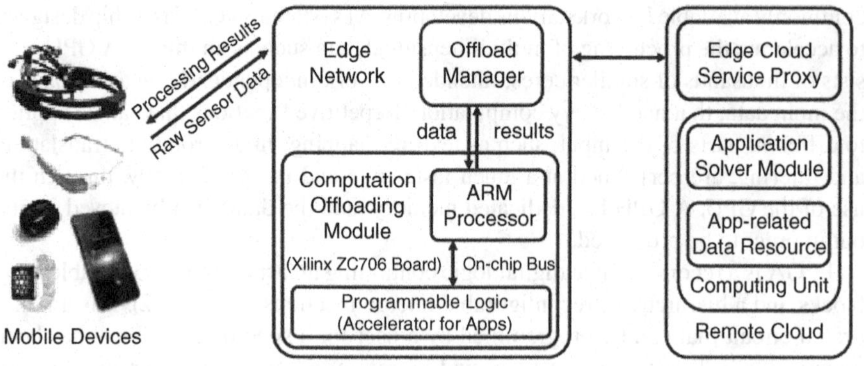

FIGURE 1.3 FPGA network systematic logic.

A typical field programmable gate array design logic is explained in Figure 1.1, and it explains how each task is interdependent on each other, including verification.

Design Entry: When text sends signal for design implementation, it had to route through behavioral simulation loop for verification. Even in design implementation stage, functional simulation, static timing analysis, and back annotation are essential before planning for programming in following stages:

- Text entry
- HDL
- Very high speed integrated hardware description language (VHDL), Verilog, System C
- Hierarchical instantiation of blocks

A typical FPGA may also have dedicated memory blocks, digital clock manager, IO banks, and several other features, which vary across different models. While a GPU is designed to perform efficiently, with similar threads on different subsets of the input, an FPGA is designed to parallelize sequential serial processing of the same program from mobile device to Edge network to Edge cloud service; a proxy logic flow diagram is explained in Figure 1.3.

1.4 BIG DATA AND DATA ANALYTICS FOR OIL WELLS

The concept of "Big Data," defined as an increasing volume, variety, and velocity of data, is familiar to the geophysical survey in the oil and gas industry. The industry generates a huge quantity of data, whether it is 3D seismic surveys, intelligent drilling well data, reservoir simulation data, production analysis data, or the monitoring of production surface facilities (pressures, flow rates, temperatures, etc.). The ability of oil man to generate, collect, and store these data has continually increased, but this has brought with it problems with managing and analyzing such vast quantities of information. Analysis of such data has been a major area of focus and innovation in the recent years within the oil and gas industry (as well as many other industries) with a view to potential improvements in exploration and production efficiency and safety.

Big exploration and production projects had covered applications in oil and gas exploration, drilling, reservoir engineering and production engineering, transportation, and marketing. Some examples include the analysis of huge micro-seismic datasets using the Hadoop-1 platform to model fracture propagation during hydraulic fracturing, the utilization of Big Data to optimize steam-assisted gravity drainage and cyclic steam flood operations on a heavy oil reservoirs by analyzing the data from fields, and optimization of the performance of electric submersible pumps by using data from huge logs.data.

Various oil and gas service providers are early adopters of Big Data analysis. GE Digital has developed "Predix," a digital platform that can be used to produce "Digital Twins"—software representations of a physical asset. The application's machine learning algorithms are able to process a huge amount of data collected by sensors, such as equipment or parts performance, environmental data, and weather conditions related to a processing facility. The algorithms then compare these against the ideal performance data contained in the database to search for discrepancies between the current and ideal state. If such discrepancies are identified, the application is triggered to send an alert to technicians, who in turn conduct preventative maintenance or part replacement.

1.4.1 Long-Reach Digital Data

Big Data are dipping into the digital oil well to leverage the latest advances in intelligent visual monitoring technology to give oil companies the ability to limit remote site visits, while reducing operational expenditure and mitigating safety and environmental risks.

When there is an alert—visual or from another sensor—operators can view live images and video to validate the process parameters alarm, have better awareness of the environment, and repair needs and any risks before deploying crews at site.

Increasingly sophisticated machine learning can "train" the system to detect and distinguish increasingly specific objects and events—for example, discriminating between a person or vehicle and an animal, or a change in cadence of a pump to provide an early warning of pending breakdown or indicating maintenance is required for the machine.

It's an important technology because it really filters out the noise and gives people more concise information. Its effectiveness is due to advancements in AI, cloud computing, and computing power to run the algorithms, and all these factors combine together—a combination of people and technology—to make it really powerful today.

The fully managed oilfields operate edge devices, including cameras, modems, processors, digital control system, and local storage, at scale, thereby reducing routine site visits by approximately 50%.

Visual evidence is also valuable for efficiently auditing third-party contractors for health and safety compliance, incidents, and invoice accuracy. Some companies used Osprey Reach to create reports showing contract workers arriving and departing, with time-stamped images, resulting in a 30% invoice reduction based on inaccurate billing. The "pictures don't lie" concept also shortens the time to resolution dramatically.

Thermal imaging can detect large methane leaks such as from storage tanks, process vessels, and pipelines. Osprey has launched two new projects looking to take that capacity to new levels in order to detect much lower volume and infrequent fugitive emissions. One project uses computer vision to automate and increase the accuracy of detecting and quantifying methane leaks, and the other project investigates capturing and integrating audio data to provide 24/7 sight and sound monitoring and analytics as part of the company's managed service offering.

1.5 DATA FILTERING AND CONDITIONING INTERFERENCE FROM SIGNALS

Many analog signals—temperature, pressure, strain, and so on—can benefit from some type of conditioning to improve the quality of measurement. Filtering is a type of conditioning that removes interference from system signals received from operating system.

The data acquisition interface might filter the signal (analog filtering); alternatively, it could use analysis software for filtering. The focus is primarily on analog filtering in hardware. For this purpose, a filter is a device that removes undesired signals according to their frequency. If the frequency spectra of signals and interference are sufficiently different, filtering can be very effective.

Filters can be built in a number of different technologies. The same transfer function can be realized in several different ways, i.e., the mathematical properties of the filter are the same but the physical properties are quite different. Often, the components in different technologies are directly analogous to each other, and, they fulfill the same role in their respective filters. For instance, the resistors, inductors, and capacitors of electronics correspond, respectively, to dampers, masses, and springs in mechanics. Likewise, there are corresponding components in distributed-element filters.

Electronic filters were originally entirely passive consisting of resistance, inductance, and capacitance. Active technology makes design easier and opens up new possibilities in filter specifications.

Digital filters operate on signals represented in digital form. The essence of a digital filter is that it directly implements a mathematical algorithm, corresponding to the desired filter transfer function, in its programming or microcode.

Mechanical filters are built out of mechanical components. In the vast majority of cases, they are used to process an electronic signal, and transducers are provided to convert this signal to and from a mechanical vibration. However, examples do exist of filters that have been designed for operation entirely in the mechanical domain. Distributed-element filters are constructed out of components made from small pieces of transmission line or other distributed elements. Distributed-element filters have structures both that directly correspond to the lumped elements of electronic filters and that are unique to this class of technology.

Waveguide filters consist of waveguide components or components inserted in the waveguide. Waveguides are a class of transmission line, and many structures of distributed-element filters, for instance, the stub, can also be implemented in waveguides.

Optical filters were originally developed for purposes other than signal processing such as lighting and photography. With the rise of optical fiber technology, however, optical filters increasingly find signal processing applications, and signal processing filter terminology, such as long pass and short pass, is entering the field.

Transversal filter, or delay line filter, works by summing copies of the input after various time delays. This can be implemented with various technologies including analog delay lines, active circuitry, charge couple device (CCD) delay lines, or entirely in the digital domain.

1.5.1 Amplifying Transactions on the Chain

Some companies have been making waves through the industry as a pioneer in applying block chain—the distributed ledger technology initially made cyber currencies possible to the oil patch. A Calgary-based company accomplished the world's first proof-of-concept energy royalty transaction on block chain using Royalty Ledger, the company's royalty application on its Energy Block Exchange (EBX), in February 2018.

EBX is a block chain business network platform for applications that the industries said will drastically reduce general and administrative costs, mitigate costly data and contract disputes, and allow for near-instant transactions in the energy sector.

Because it uses an immutable and transparent ledger, block chain removes the need to trust the parties between business transactions—trust is built into the platform. "What Bitcoin proved was that can transact with strangers and people that it didn't trust and it could derive trust from the system," said by one industry expert. Royalty Ledger, for example, uses a mutually agreed-upon contract as the basis of its ongoing payment calculations. After both sides agree on contractual terms, the software's business logic—based on current and historical legal conventions and precedents—automatically triggers the transaction on fulfillment of the applicable terms of the contract.

Block chain is the logical culmination of a process that started with the alignment of data first to solve data discrepancies and disputes between company departments (data silos), then between vendors (platforms), and finally between companies (contracted counterparties and partners):

- The interpretations of the contracts and a common interpretation of the calculation of how it can execute the transaction.
- All in the middle between the counterparties in a shared space. The benefit that is emerging there is the concept of frictionless transactions. It has proven that it can takc a 90-day dispute-laden process and reduce it to 90 seconds if the work to really share the facts, and smart contracting distributed ledgers allow us to do that.

An industry expert mentioned that traditionally very protective of their data, oil and gas companies were initially reluctant to consider any kind of sharing of that data. That point began to change with the downturn of 2014.

Technology partners include Amazon Web Services and recognized enterprise block chain software firm R3. Among other things, bringing them on helped in terms of security, scalability, and reliability—bringing immediate credibility to the technology application.

1.5.2 Augmenting Collaboration

A typical Panoptic model was created to address the significant amount of overspend required to rework projects in the construction phase, which accounts for up to 30% of overall project expenditure. Errors in the original design that were not caught during the initial design review and planning phase are the main reasons for rework. Unlike traditional tools that display 3D models on a 2D screen and lack the context needed to fully comprehend scale differences, Panoptica accurately conveys the true sense of size and space in relation to real-world surroundings. During engineering design review meetings, teams are able to walk up to, around, duck underneath, and engage with 3D models using natural human interactions as if the models were physically present.

1.6 CASE-BASED REASONING

Fuzzy logic is an AI-based technique used for prediction and reasoning when information is unreliable and/or incomplete. This makes it particularly useful for filling information gaps during oil and gas reservoir characterization. It has also become increasingly common in other engineering control applications in the energy industry where input parameters are highly variable. These uses include enhanced recovery, well stimulation, and infill drilling.

Although an increasing number of oil and gas companies are embracing AI-based technology, the industry as a whole still lags many others when it comes (Figure 1.4) to fully leverage its data. Breaking down information silos and making data more available to key decision-makers across various disciplines will be integral to improving this situation in coming years. In the short term, however, the use of machine learning, predictive analytics, and other data-driven solutions will continue to help companies improve efficiency and remain profitable in the current economic environment.

Although their hand may have been forced by low cost for end products, those who have made the investment in new technologies may be able to emerge from the downturn stronger than ever. Figure 1.5 explains global digital oilfield market trend with respect to various groups like process, solution, application, and region, and these are explained in the flow chart.

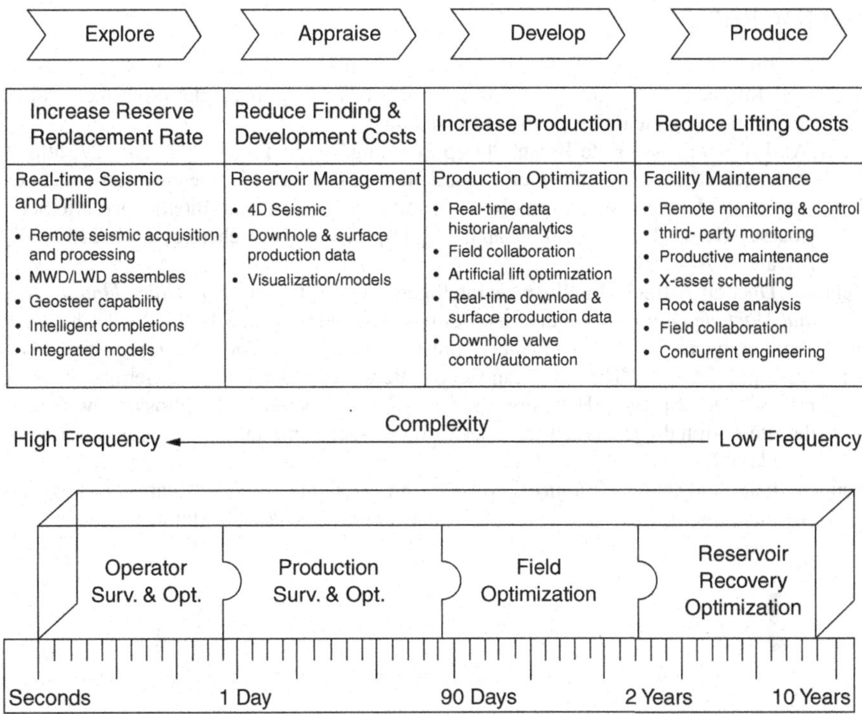

FIGURE 1.4 Flow diagram of different stages from oil exploration to production.

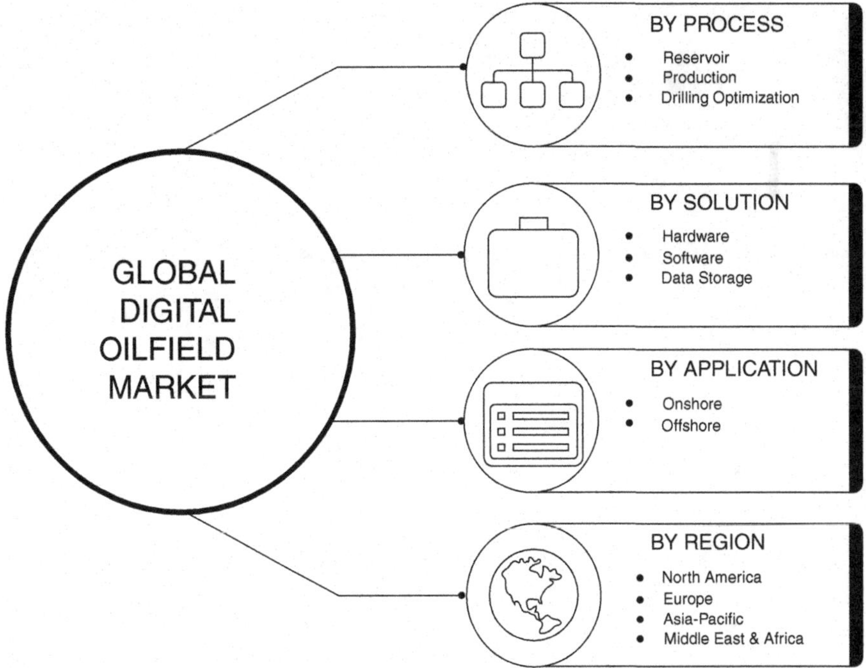

FIGURE 1.5 Role of global digital oilfield market.

BIBLIOGRAPHY

Bhaskar, Jitu. "How to Develop an eLearning Website and App Like Udemy." Semi Dot Infotech Data Group. 2020 September 7. https://semidotinfotech.com/blog/how-to-develop-elearning-platform-like-udemy/.

Gillis, A., Ed Burns, and Kate Brush. "Deep Learning." Tech Target. 2023 July 27. https://www.techtarget.com/searchenterpriseai/definition/deep-learning-deep-neural-network.

Kennesaw State University. "Advanced A. I Answers - Artificial Intelligence Method." Course Hero, USA. 2017 March 4. https://www.coursehero.com/file/21590867/Advanced-AI-Answers/.

Mehrotra, Dheeraj. Artificial Intelligence: Intelligence Types. In *Basics of Artificial Intelligence and Machine Learning*, Kindle First Edition, Notion Press, New Delhi. 2019. Chapter 3.

Taulli, Tom. *Artificial Intelligence Basics*, Apress Publisher, New York. 2019 August. 31–57.

Von Christopher, Alessi. "The European Central Bank Tests Set to Reveal German Banking Faults." Der Spiegel. Hamburg, Germany. 2013 October 23. https://www.spiegel.de/international/europe/ecb-tests-set-to-reveal-german-banking-faults-and-political-agenda-a-929553.html.

Webber, Sheila, Jod Detjens, Tammy L. Maclean, and Dominic Thomas. "Team Challenges: Is Artificial Intelligence the Solution?" Business Horizons. 2019 September.

2 Artificial Intelligence Agent System

2.1 INTRODUCTION

Digital technology that began with Internet and mobile technologies plunges corporations into opening their stores in cloud and web and mobilizing e-government initiatives, and financial institutions into presenting themselves in tablets, mobile phones, and social media. The massive transformation of business processes reengineering shifted the Industrial Age toward the digital age with the help of e-business environments. Huge information getting bigger and bigger every day led the business environment to analyze big data and react simultaneously with customer relationship management (CRM) systems. Although the digital age, together with other sciences like mechatronics, nanotechnology, and genetics, is a step for "Space Economics," some other signs of progress will change business and economics directly or indirectly more than other developments. This progress is named Robotics and Artificial Intelligence. "Industrial Age" was started in the UK by car makers. The Industrial Age developments affected the production factors, i.e., capital, entrepreneurship, workforce, and land. Due to these effects, mechanization, lifestyle, education, finance, and management have all changed. White collars and administration have joined the agenda to solve new issues and problems, which supplements the higher education needs due to the information, decisions, and quality of the workforce needed. The arguments suppose that robots and artificial intelligence (AI) might be positioned even in managerial positions, and first comers begin to prove that assertion. Minimizing the cost of delivering internal documents between head office and local office by drone's technology would be another way to reduce the costs. On the contrary, drone operators will be hired by financial institutions and companies; Amazon.com is an excellent example for this. But the secondary effects of using drones for commercial purposes are the main deliberation points. Robotics science with Space Economics tries to find solutions for the commercial use of drones, spaceships, and rockets due to existing laws and aviation regulations that only allow governments to use, launch, and fly rockets or spacecrafts.

Deliberation on the security of data and information and the life privacy of individuals are also other issues that humanity should tackle. One of the new systems is transmitted by a voice recognition system.

AI systems also consist of an agent and its environment. An agent such as a human or a robot identifies the environment through sensors and effectors. It uses a method called search and pattern matching, where the computer is instructed to search its

DOI: 10.1201/9781003307723-2

knowledge base based on the match found and if specific conditions are met to solve a problem. Various agent-based systems and AI that are closely interacting and coordinating with the new concept of an autonomous entity that could plan itself are discussed below.

2.1.1 Multi-Agent Systems

Agent-based systems and AI are closely related so far as interaction, contribution, and dependency are concerned. The traditional AI problem-solving method emerged around the 1990s with a new concept of an autonomous entity that could plan itself. Its course of action is routine work or situation initiated by an external agency to create an environment of its kind using adaptation, self-organization, and learning. The autonomous entity recognized as agent has reactivity, pro-activeness, and societal apart from autonomy. Autonomy is the in-built and fundamental characteristic of agent, as logic is to an algorithm. The agent may be alone like user's interface agents or software secretaries, and it may be like the master slaves model, i.e., one master having the most significant capability connected with low-power agents; it may be like a multi-agent paradigm in which homogenous or heterogeneous kind of agent work in coordination and cooperation.

2.1.2 Autonomy

It is the ability to perform largely independent of human intervention. In another way, it encapsulates some states not accessible or apparent to other agents and uses a decision-based approach.

2.1.3 Reactivity

It is the ability to perceive the surroundings and respond on time to events occurring in them. The surrounding may be the physical world, a user via the Internet, or any one of these. By perceiving the surrounding, the agent should be adaptive enough to accommodate its changes and respond as and when needed.

2.1.4 Reactiveness

On the situation's demand to take the initiative, the agents display a goal-directed behavior by taking initiatives despite responding to their surroundings.

'Agents' are personified as human beings working in society. Agents working in multi-agent surroundings exhibit the characteristic of human behavior in an organization, such as coordination, cooperation, and communication.

2.1.5 Collaborative Agents

Many agents work in a collaborative exhibiting, coordinating, and cooperating manner among themselves. Each agent has its autonomy as a dominant factor rather than dependent and working in a constrained multi-agent environment. The agents reach a consensus or a shared agreement point through negotiation. The agents will be assigned a particular type of job, and they will coordinate, cooperate, and communicate with each other section dealing with the blackboard model.

2.1.6 Interface Agents

It is a secretariat level of assistance offered to a user collaboratively. The critical point is to perform their task. The interaction with other agents is only up to the level of advice.

2.1.7 Mobile Agents

Software processes incapable of roaming in wild area networks such as www perform their tasks on behalf of their owner by interacting with other foreign hosts. After completion of the job, they return to their respective "home." The position may be of different kinds—oil transportation by tankers.

2.1.8 Information Agents

Nowadays, the multiple-agent paradigm is being used extensively in information systems, particularly in distributed systems such as the Internet. A dynamic cooperative environment performs the manipulation, retrieval, and management of information. The softbot Internet is an example of an information agent.

2.1.9 Reactive Agents

The reactive agents work in a simulating response paradigm and environments such as coordination and organizational attributes. These agents lack the explicitness in their goal.

2.1.10 Hybrid Agents

Rational agents, knowledge agents, and socially responsible agents are the kinds of agents classified based on quite a different level of abstraction. The contribution of these agents or any other type of agent makes hybridization products and works collectively on a particular goal.

2.1.11 Heterogeneous Agents

The agents with different characteristics and capabilities are integrated to form a holistic model. Their working problems are manifold when an integrated model of such combination is implemented, particularly conflict resolution.

2.1.12 Responsible Agents

Five attributes of the responsible agents are given based on their nomenclature:

The system: The entity is to be described below:

2.1.12.1 System

The multi-agent systems basically involve two levels of manifestation. One is known as knowledge level, and another is known as social level. The agents have five different attributes, as mentioned above. At the social level, the actor is responsible for society. The autonomous agents interact with each other depending on their choice and desire. Their choice affects other agents individually, and further depending also affects their activities.

2.1.12.2 Components

The responsible society is made of components as members, environment, interaction means, and goal. Socially responsible agents are the members who perform the problem-solving responsibility. Depending on the representation in knowledge level and society levels, these are termed agents and members. These problem-solving entities can be viewed at different prospective, granularity, and abstraction levels. The entities have their workplace, and the context where these are active is known as the environment. It may be a factory, plant, office, organization, or the Internet. The agents interact with other agents and the environment differently, i.e., through the Internet in information processing, blackboard for control, and such actions where the mutual benefit is more significant than collective loss. The responsible agent may exhibit a different type of behavior. Society may gain in total when members perform such action, which may only benefit them and not the organization. Individual steps are those that contribute net benefit to the individual but not to society. When the members are concerned and the community gets help from their actions, it is called the dividend. The move is not helpful when both individuals and society are not benefited.

2.1.12.3 Medium

A member possesses the ability to obtain the behavior specified by the principle of social rationality based on the terms of acquaintance, rights, and duties. In this case, acquaintance referred to the point that a member should be aware of the activities of the other members of the society. Society should also be mindful of the actions of the members. Under certain circumstances, a member should try to influence the goal of the other members of the society, and similarly, the community should be able to control the members.

2.1.12.4 Compositional Laws

The multi-agent system can be organized in different ways or topologies, i.e., matrix, hierarchical, circular, or linear, depending on the characteristics of the problem domain. As far as the problem of inheritance is concerned, hierarchical topology is the most suitable approach, and where the mutual relationship among the active entities, dimensional representation of the activities, and communications is concerned, matrix topology is the preferred approach.

2.1.12.5 Behavioral Laws

All the actions performed by the members should contribute a net benefit to society by the law:

$$\sum_{j \in S} Nj\ (M, a) > 0,$$

where S is the society under study
j is the number of members in the society under study
M is a responsible member of S
A (M) is the set of actions to perform
A is an element of A(M)
Nj (M, a) is the net benefit for member j.

2.1.13 AI Components in Multi-Agent Systems

With its two components, knowledge representation (KR) and reasoning, AI can fully model knowledge and logic involved in the multi-agent paradigm.

2.1.13.1 Knowledge Representation

Three knowledge structures are required in a plan-based system for responding. An action specification represents a mental knowledge level for reasoning about the plan construction process, an object knowledge level for a reason about domain objects, and a third level for learning the logical consequences of plan actions.

2.1.13.2 Reasoning

A plan-based system adapts the automatic reasoning process such as non-monotonic default logic or circumscriptions. These reasoning models have hindered the execution time. But the environment changes to accommodate the changes so rapidly that to accommodate the changes in the reasoning methods become unmanageable. Therefore, altogether a different ball game for the reactive agent is needed.

2.1.13.3 Common Sense Reasoning

In most applications, common sense reasoning involves the motion as time and other entities are modeled as specific to the application domain informally.

That is why it is incapable of modeling the reasoning process in the intelligent agent system.

2.1.13.4 Model Reasoning

Model reasoning system cannot be applicable for the simple reason that an agent believes that all the logical belief statements cannot be assumed to hold in other belief statements.

2.1.13.5 Temporal Reasoning

Reasoning about time still presents a significant challenge for AI and, therefore, to agent-based systems, although attempts have been made to utilize temporal logic in AOP.

2.1.14 AOP Components and Modalities

The latent attributes of an agent depend on belief, desire, choice, and motivations. Various authors have proposed different combinations of these factors. In agent-orientated programming, thought is common to other modalities, but commitment is related to obligation, choice, and decision. The capabilities are not a behavioral term but a physical measure; in control, it is in conjunction with the behavioral qualities to form a modality.

2.1.15 Blackboard Architecture

The blackboard model is an architecture developed by Hayes Roth to control interference mechanisms in the presence of distributed, multi-experts, basically, the distributed reasoning in particular and distributed AI in general, and the multi-agent paradigm. The problem is divided into the hierarchy solution, each at a different level of abstraction.

2.2 USER INTERFACE

A user interface that involves some aspects of AI is known as an intelligent user interface (intelligent UI, IUI, or sometimes interface agent). Generally, an IUI involves the computer side having sophisticated knowledge of the domain, including a user model. The computer–human interface design involves computer graphics, sound systems, speech synthesis, speech recognition, and haptics. These aspects of the interface have already been developed and designed with established methodologies. But the mode of integration of these functional entities for human perception, cognition, and factors transforms this bodily affair to the minimum, which involves the physics and physiology of perception rather than computational model and engineering design. The interface should have the following characteristics:

- This should be robust enough, complementary to efficient perception, and instructive for the novice.
- This should be facilitating in recovery from the manipulative mistake and cognitive.
- This should be suggestive for corrective actions and helpful in the diagnosis of errors.
- A UI is a method by which the user and the computer exchange information and instructions. Generally, there are three types of UI—command line, menu-driven, and graphical user interface (GUI). A typical example of a hardware device with a UI is a remote control. Other devices also have a UI, such as digital cameras, audio mixing consoles, and stereo systems.

2.2.1 Computer Interface Specification

The human–computer interface is one of the essential knowledge base systems, an interconnected system for designing developing is user-friendly interface components for humans and computers. All these functions or components are classified into the following interface, which is necessary for designing an interactive system:

- Developer
- End-user
- Data
- System
- Cost and vendor related

This interface should have considered the factors such as robustness, user, familiarity, security friendliness, and compatibility. Few human factors should be taken into consideration while designing an interactive system, which are given as follows:

- The system should be adaptable.
- The system should have a high degree of flexibility.
- The system user's language should be the same.
- The system should have a scope for multi-modal communication.

2.2.2 Human Factors

The primary interaction between the user and the system as a dialogue session should be uniform, short, and perfect as far as language is concerned. The uniformity should be maintained whether it is natural language, command language, or simply a binary (answer in "Y"/"N" only).

2.2.3 Types of Human User Interface

The following are the various types of a human UI system:

- Command line (CLI)
- Graphical user (GUI)
- Menu-driven (MDI)
- Form-based (FBI)
- Natural language (NLI)

Computer interfaces can be designed for either hardware or software; most of the interfaces are a combination of both. The AIS is a part of service availability interfaces of the service availability forum.

2.2.4 Classification of AIS Services

AIS consists of 12 services and 2 frameworks. The services are classified into three functional groups as AIS Platform Services, essential AIS Management Services, and General AIS Utilities.

Initially, the application programming interfaces (APIs) had been defined in the C programming language only. The Java mapping of the different service APIs was released incrementally.

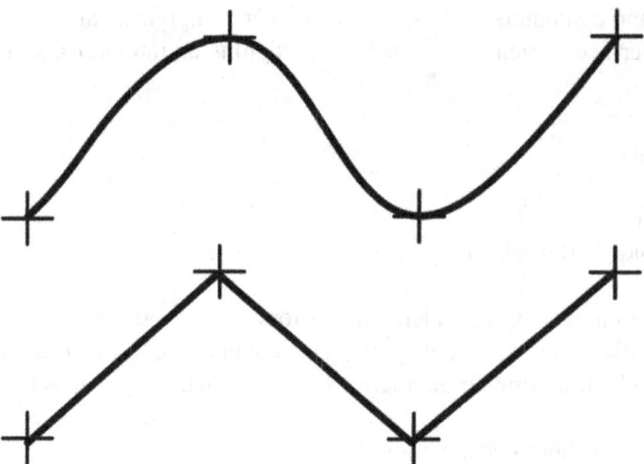

FIGURE 2.1 The camera sample speed of the wheel spoke movement plot.

Various services and frameworks of the interface specifications are designed to be modular and independent of one another to a certain degree. This process permits a system providing only AIS and no HPI to exist and vice versa.

The only necessary architectural dependency is the dependence on the Cluster Membership Service (contract lifecycle management CLM). All AIS Services, except the Platform Management Service (product lifecycle management PLM) and the Timer Service, depend on CLM.

All AIS Services are expected to use the AIS Management Services to expose their administrative interfaces, configuration, and runtime management information (Figure 2.1).

The PLM provides a new logical view of the system's hardware and the low-level software. Low-level software, in this sense, consists of the operating system and virtualization layers that provide execution environments for all kinds of software.

The primary logical entities incorporated by PLM are as follows:

- Virtualized architectures in the PLM Information Model.
- Hardware Element (HE): An HE is a logical entity that represents any type of hardware entity; this may be, for instance, a chassis, a CPU blade, or an I/O device. Typically, all field-replaceable units are modeled as HEs.
- Execution Environment (EE): An EE is a logical entity representing an environment capable of running software programs. In case a CPU blade or an SMP machine runs a single operating system instance modeled as EE.

Various virtualization architectures are generally supported. In this case, PLM maintains the state of these entities in the information model and supports the means to control them and track any condition changes. To fulfill these assignments for HEs, the PLM Service typically uses HPI. In the case of EEs, PLM is in charge of retrieving all necessary information related to the health of the operating system and any of the available virtualization layers.

The CLM provides applications with membership information about the nodes that have been administratively configured in the cluster configuration (these nodes are also called cluster nodes or configured nodes) and is core to any clustered system. A cluster consists of this set of configured nodes, each with a unique node name.

Two types of logical entities implemented by the CLM are as follows:

- Cluster: Represents the cluster itself, which is the parent object of the cluster node objects.
- Cluster Node: Represents a configured cluster node.

The CLM generally provides APIs to retrieve the present cluster membership information and to track membership changes (e.g., node leave and, node join). All cluster-comprehensive AIS Services must use the CLM track API to determine the membership.

2.3 KNOWLEDGE-BASED SYSTEMS

The primary system contains the component of a knowledge-based system (KBS). The essential elements of KBS are KR, interference, and interface. A KBS should include a KR scheme and the exact constituents of interference such as reasoning, control, and uncertainties measurement and interfaces.

2.3.1 Structure of an Expert System

An expert system consists of (1) knowledge base, (2) inference engine, (3) knowledge acquisition, (4) explanation module, and (5) user's interface.

2.4 KNOWLEDGE DISCOVERY OF DATA AND WEB MINING

Data mining is the collection of techniques for extracting knowledge from an extensive database. Knowledge discovery of data (KDD) is also manifested as knowledge or data mining. Knowledge discovery and data mining concepts are interleaved and interconnected. Sometimes, it is challenging to identify in either process. The method of data mining, data forming, application, data, and KDD packages have definite salient characters.

2.4.1 KDD

Knowledge discovery of data is related to classification and clustering of support data transformation and visualization. Medical decision-making, marketing, and patient care in the segment of oil and industry are given below.

2.4.2 Business

It consists of domain experts, retail marketing, consumer care, and consumer relationship management. Various business components like customer classification and vendor approval yield production are considered based on ranking importance that affects supply chain management.

2.4.3 Industries

In the process of manufacturing, there are various types of design software like CAD, CAM, CIM, etc.

2.4.4 Web Mining

It gives importance to the data mining process beyond the traditional analysis of structured data. It is mainly related to extraction information from a considerable amount of data available in the free text, images, sound, and video.

2.4.5 KDD Methods

It consists of a complete and clear understanding of the end user's objective. Knowledge of the proper application domain is the first and foremost step in this process and further target and data collection.

2.4.6 Data Cleaning and Pre-Processing

The raw data need to be conditioned according to the required specific type of refined data, which requires increasing signal to noise ratio by filtering.

2.4.7 Data Reduction and Transformation

Various algorithms are available in the software market for data reduction and compression, mainly in image processing for better visualization on websites.

2.4.8 Data Mining Task

There are several studies that can be performed to achieve the goal, which is carried out by processes like classification, clustering, and regression.

2.4.9 Solution Strategy

Selection of algorithm, solution strategy, and methods of forecasting pattern of interest in the data.

2.4.10 Interpretation

Rules are induced or generated based on the mined data pattern.

2.4.11 Evaluation

The applied method for evaluation of patterned mine data concerning the goal and measure its performance.

2.5 WEB TECHNOLOGY, SEMANTIC WEB, AND KNOWLEDGE MANAGEMENT

It is essential to have languages, reasoning, and management technique for ontology. The ontology languages used for properly expressing the ontologies should have good expressivity, web compliance, and support reasoning. These reasoning processes should be stable, scalable interference machines. This reasoning process should be fault-tolerant. However, the mapping, merging, and alignment of ontologies are provided by the ontology integration technique. Software like Fact ++ and Racer deploy their reasoning mechanism based on the applications to the ontologies and their expressive power. Descriptive language (DL) is an inference engine for OWL lite and OWL DL but not for OWL full. It is challenging to have the formal property for most of the inference engines of OWL family languages implemented for the semantic web.

2.5.1 Semantic Web Services (SWS) Framework

The framework for SWS includes the following:

- SWSA: Semantic web services initiative architecture is built on the W3C. This architecture group design is based on the version of Tim Berners-Lee on Semantic Web OWL-s web ontology language for services (OWL-S) consortium.
- WSMO: Web Service Modeling Ontology.
- WTEOR-S: Managing end-to-end operations semantics.

These protocols interchange between the interacting agents. Agents interpret and reason semantic descriptions when deploying Semantic Web Services (SWS). There are five classes of SW agents in active service: discovery service, engagement service, process enactment, management support services, and quantity of service.

2.5.2 Knowledge Management

Different persons in different contexts have defined the KM. KM is the discipline that provides the strategy process and technology to share and leverage information, and expertise that will increase the level of understanding to make a proper decision. The integration of information technology and business processes facilitates the efforts to capture, store, and deploy knowledge. The integration leads to the generation of the following strategy:

The business process consists of knowledge and learning capabilities.

The recommended corrective actions are given below:

- Adaptive enterprise guiding.
- Emphasis on effectiveness versus efficiency.
- Emphasis on transformation management.

Knowledge management framework covers the following function:

This addresses the opportunities for KW assets, effects of their use—a current obstacle to its use, and estimation of the increase in KW.

- It specifies the actions that forbade usability and added value of KW.
- It requires the planning actions to use KW assets to action and monitor activity.
- It was reviewing of KW to ensure added value.
- It formed to know the desired added value and the strategy to maintain KW assets and create new opportunities.

2.6 DATA FILTERING FROM SIGNALS

Many analog signals—temperature, pressure, strain—can benefit from some type of filtering to improve measurement quality. Filtering is a type of conditioning that removes interference from the signals.

The data acquisition interface generally filters the signal (analog filtering); alternatively, it could use analysis software for filtering. Here it is focusing primarily on analog filtering in hardware.

2.6.1 Data Analysis Adds Value to Content Filtering

AI-powered content filtering can do much more than analyzing websites for appropriateness or picking up red flags requiring action. It can also provide meaningful data for decision-making.

For this purpose, a filter is a device that removes various undesired signals based on their frequency. If this frequency spectrum of signals and interference are sufficiently different, filtering can be very effective.

2.6.2 Choosing the Right Type of Filter

2.6.2.1 High-Pass Filter

A high-pass filter might be helpful when a very low-level transducer signal is superimposed on a large DC output voltage. This attenuates (removes) low frequencies. Using a cut-off frequency of, say, 5 Hz will eliminate the DC voltage, which has a frequency of zero.

A high-pass filter will remove "drift." It can be a problem with biological and chemical signals, but not with modern electronic signals.

2.6.2.2 Low-Pass Filter

This generally used a low-pass filter. It passes through the lower frequencies and attenuates the higher frequencies. Select the cut-off frequency compatible with the unwanted frequencies, the frequencies present in the signal you are measuring, and the analog-to-digital converter's sampling rate.

There are several other ways of removing high-frequency noise from the signals. The amplifier itself has a high-frequency cut-off—an integration with an A-D converter which will behave as a low-pass filter. Keeping signal wires as short as possible, wrapped together or shielded wires, keeping away from all electrical equipment, and using differential inputs will help reduce interference.

2.6.3 Anti-Aliasing Filter

When recording a signal waveform, a computerized SCADA takes readings from (samples) the signal and interpolates for whatever the signal is doing between assignments. It combines up the dots to make a waveform. Too few drops can produce a wrong picture, showing a waveform with much too low a frequency. This is called Aliasing.

For example, aliasing occurs when stagecoach wheels on a film go back to the coach traveling forward. The camera samples the scene 24 times per second. This sampling rate is prolonged for the speed of the spokes of the wheel. Instead of each frame showing the spokes a little bit further round, each spoke moves to just short of the original position of the said in front, giving the appearance of rotating backward. It will be around 10–20 times the highest frequency component of the actual signal in practice.

The angular waveform is generated when the signal is sampled at twice its highest frequency. For a more accurate representation, the sample is produced 5–10 times at this rate. Well, if there is high-frequency interference and then sample is according to the lower frequency of the signal, it will alias the interference and make it look like a part of the signal that is trying to measure. It can also remove the interference utilizing an anti-aliasing filter, which is a type of low-pass filter.

An anti-aliasing filter has a sharper cut-off than an average low-pass filter. It is specified according to the system's sampling rate, and hence, there should be one filter for every input signal. In data acquisition using a 12-bit analog-to-digital converter, attenuation of –78 dB will get rid of signs that the converter can't resolve—some guidelines for selecting the filter's cut-off frequency are given in Table 2.1.

TABLE 2.1
Filter Cut-Off Frequency Guideline

Signal Type	Cut-Off Frequency Criterion
Pulsed DC	Rate of change (V/sec/1V)
Sinusoid	1/period
Complex periodic	20/fundamental period
Single events	1/pulse width

2.6.4 Band-Pass and Band-Stop Filter

A band-pass filter permits only a pre-defined frequency band through, while a band-stop filter does the reverse—stopping a pre-defined frequency band. A band-pass filter may be helpful when the signals are all at one frequency, and a band-stop filter removes mains interference. For example, if the signal is 0–200 Hz, it could use a band-stop filter to remove the 50 or 60 Hz mains band. This would slightly distort when it measured signal but by much less than letting through the mains interference.

2.6.5 Post-Collection Filtering

Provided the signal has been accurately recorded, particular aliasing effects have been avoided, which may view the signal waveform on the computer display system and apply new software filtering if required. We use software for system development projects to review and post-process the acquired waveforms, allowing the design of standard digital filter types. This way, the frequency of response means that complex design-related calculations carried out on the recorded waveforms (e.g., Head Injury Criteria) can be compared between various test laboratories.

BIBLIOGRAPHY

Carbonell, Jaime R., and Allan M. Collins. "Natural Semantics in Artificial Intelligence." *Proceedings of the Third International Joint Conference on Artificial Intelligence*, 1973 September. 3451.

Markovitch, Shaul, and P. D. Scot. Information Filtering: Selection *Mechanisms in Learning*, Machine Learning. 1993. 10: 113–151.

Maybury, Mark T., and Wahlster Wolfgang. *Intelligent User Interfaces*, Morgan Kaufmann Publisher Inc., San Francisco, CA. 1998. 443–516.

Mishra, R. B. *Artificial Intelligence*, Second Edition, PHI Learning Private Ltd., New Delhi. 2013. 214321.

Rajaraman, V., and Neeharika Adabala. *Fundamentals of Computers*, Sixth Edition, PHI Learning Private Ltd., New Delhi. 2015. 9–32.

Su, Xiaoyuan and M. Khoshgoftaar Taghi. A Survey of Collaborative Filtering Techniques, *Advances in Artificial Intelligence*. 2009. 12: Article ID 421426. https://doi.org/10.1155/2009/421425

Zhang, Zili, and Chengqi Zhang. *Agent-Based Hybrid Intelligent Systems*, Springer Link, New York. LNCS vol-2930, Chapter 4. 2004. 43–55

3 Artificial Intelligence and Automatic Control

3.1 INTRODUCTION

Artificial intelligence (AI) is the science that enables computers and machines to learn, judge, and use their reasons. As technologies are becoming more complex, the demand for AI is increasing because of its ability to solve complex problems with limited human resources and expertise and within a limited time. AI adopts the abilities to equip the technical knowledge and amplify expertise to learn and deploy new methods and applications. There is a significant breakthrough in image recognition using machine learning and advances in big data and Graphics Processing Units, which helped AI grow faster. AI systems consist of an agent and its environment. An agent such as a human or a robot identifies the environment through sensors and effectors. It uses a method called search and pattern matching, where the computer is instructed to search its knowledge based on the match found and if specific conditions are met to solve a problem. In education, AI can provide auto-grading, help students learn by supporting their needs, and help them stay on track. AI has helped lawyers go through thousands of large legal documents effectively and accurately in the law industry, which is generally tedious. Industrial robots have made manufacturing way easy and efficient than it used to be a few years back.

The most important issue that can lead to a rethinking of present achievements and theory and practice of AI is the sharp tendency in possibilities of using computer system including hardware implementation of logical and other means of use AI. The term "intellectual control system" refers to any combination of hardware and software, which is joined by a general information process, operating autonomously or in man–machine mode, and capable of synthesizing the control goal and finding rational ways to achieve the control goal.

3.2 CONCEPT OF AUTOMATION

The use of automation began to increase in the last decade to reduce the workforce and time. Automation has introduced a system of computers and machines and replaced a system built by combining man and machine. Highly intense and repetitive tasks have become efficient, and the product quality has also increased with automation in various industries. There are multiple types of automation; some of the popular ones are given as follows.

3.2.1 Numerical Control

Drills, 3D printing, glass cutting, etc., fall in this category where machines are programmed to execute repetitive tasks.

DOI: 10.1201/9781003307723-3

3.2.2 Computer-Aided Manufacturing

Computer software is used for this automation, such as computer-aided design, and computer-aided design and drafting.

3.2.3 Flexible Manufacturing Systems (FMS)

It is a sophisticated automation system where robots and other advanced automation tools provide flexibility and customization to the users.

3.2.4 Industrial Robots

Robots are being used for routine medical operations, welding work in the construction industry, assembly, handling materials, etc. Robots can be programmed and manipulated in three or more axes.

3.3 DIFFERENCES BETWEEN AI AND AUTOMATION

Before diving into AI in automation, it's essential to acknowledge that both these terms are used interchangeably in daily life. They are associated with physical or software robots and other machines, which allow us to work more effectively and efficiently. This can be mechanical tasks like piecing together something like a car or sending off a follow-up mail just the day after you find that your customer hasn't yet completed their order. But people also fail to realize that there are also significant differences between these two. The differences correspond to the complexity level in both systems.

There are differences in designing software or hardware capable of automatically doing things without any human intervention. On the contrary, AI is a combination of science and engineering involved in making intelligent machines. AI is about attempting to make devices mimic or even supersede human intelligence and behavior. The whole practice of automation has evolved into its current form between the first and third industrial revolutions. It involves production using automatic testing, mechanical labor, and control systems that are bound by explicit programming and rules. To ensure that the same thing becomes an AI, it stimulates like a human. In the case of automatic, it will be able to know the output using sensor readings. In the case of AI, there is always a little bit of uncertainty, just like it's there with the human brain, rational judgment and thought. AI is designed to seek patterns and learn from experience to self-select the appropriate responses according to situations.

3.4 MAJOR COMPONENTS OF AI IN AUTOMATION

The use of software to reduce human effort isn't new for the business community. On top of it, AI has opened various new possibilities both in research and industrial applications. The automation provided for a minimal range of reducing human work. But by combining AI with automation will minimize human effort and remove the need for such intervention altogether. This kind of combination in AI is the automation continuum (or intelligence Robotic Process Automation).

An automation system functions using the three components of AI. So, depending on the need, they can be either combined or even used separately to allow for a fully automated response:

Natural language processing does the same on the visuals to understand human voice and text inputs. It's now possible for machines to understand the context behind the communication being carried out and then take actions based on prebuilt data and contextual variables at play.

3.5 APPLICATIONS OF AI IN AUTOMATION

AI can be used in various ways in automation control. From drones to self-driving cars, all are making use of intelligent automation. Here are some of the significant ways in which a commercial process will be able to benefit from a combo of AI and automation to face the perpetrator. There will be a camera attached to the point of sale (POS) system; it will record all types of transactions and then link them directly to the face with the details already present in the system. Say, for example, if someone commits credit card fraud, then it will become much easier to take legal action. Moreover, an intelligent system will also prevent cyberattacks by swiftly identifying the abnormal behavior of the user. The system automatically stops taking any requests and fires off an alert to the administrator in such situations. It can be made a lot easier with the help of automation. Brand marketers struggle to understand the consumer's opinion about their brand fully. So, with the use of automation, they can automate the analysis of all the present content across the internet. This can be carried out daily and will help identify critical issues.

3.6 ROLE OF AI IN SOCIAL MEDIA

Here, the user defines a set of focus words and context and understands what online users are describing in a moment, and this is done using chatbots. They have become a lot popular and that too in a brief period. It began with Apple's Siri but has been adopted by most brands now. Chatbots are programs that can understand the user's input on a contextual basis and then respond to the queries accordingly. These are used for automating customer service, sales, and marketing messages. These bots create friction during the app download process in the case of popular platforms like Facebook, messenger, etc. They feel human, too, and can reduce the burden that entrusts on customer help desks. Software testing is a very rapidly evolving field. Using a whole range of tools that are made available, all the work will be fully automated in the future. Development in automation is a long way to go for sure, but some tools can relieve developers from being engaged in menial tasks—a lot in sorting through the curricula vitae (CVs) which get submitted with them for verification. With the help of automation, this will identify the potential candidates and manage old data. These solutions upload the job application materials straight to their database when the users apply for training a human in any routine task. This will deal with employee turnover, allow gradual experience and skill development, and incur additional costs.

On the contrary, a machine must be trained only once, and then, it will improve over time and that too with absolutely no cost involved in repeat training. But an automation solution is way more foolproof, and it will not indulge in many errors. With time, it will also learn from the outputs, and hence, its efficiency will also improve further.

3.7 ENHANCEMENT OF HUMAN CONDITION WITH THE PARTICIPATION OF AI

Some of the companies do make use of AI, whether for marketing, hiring, or such. The issue that industries are facing now is the absolute lack of consumer trust. But this needs to be focused on as AI continues to overgrow. Many companies make use of AI and algorithms to make decisions for the customers. But this will lead the customers to believe that it's human. But they show the customers to think that it's the human making the decisions. This isn't the best policy because customers will have a desire to know how their information is being scrutinized, the decisions being made, and any bias in making those decisions.

Along with this, it could be asked to share the algorithms that helped determine recidivism in legal fields. Such steps will help put the power back in the hands of the consumers. The year when AI combines with other technologies like enterprise resource planning (ERP) solution, and analytics, will be the key for the success of enterprises. Convergence will give better data-driven decisions by using data from various sensors, which could improve the quality and performance. A specific team won't gain traction as all the segments of an organization need to be involved. But more than that, tech rollout, especially ones who can change the entire dynamics of an enterprise, needs to be led appropriately from the top. Along with this, leadership should ensure that making AI is a priority in the enterprise and offer retraining programs.

Employees should learn the skills so that they can run the AI programs efficiently in the organization. If the real reward is seen, the investment will have to be made in both the software and employees. This can be seen as significant openness to AI regarding automated decision-making through various businesses and organizations.

3.8 COMPARATIVE ANALYSIS AND A HIGHER LEVEL OF INTELLECTUAL CONTROL

The fact that the past decade has seen a rapid increase in the number of theoretical and applied research in fuzzy controllers; the report's primary focus is to review the significant achievements in this area. Unfortunately, even this field doesn't allow a complete review free from the authors' preferences.

3.8.1 General Problems of Control Systems Intellectualization

The successful solving of the issues to ensure the technological independence of the state in the field of civil and military purpose complex technical objects' development and application significantly depends on the effectiveness of control systems and technologies being developed. Adequate theory and control technologies are necessary, considering possible deficiency of particular (depending on application) required resources: information, timing, energy, financial, material, personnel, etc. Known accidents and disasters in transport, industry, fuel, etc., are often associated with the so-called human factor (HF), including the overwork of operators. HF usually occurs due to quality problems with the design of the control system, in particular as emergencies in controllability. Human errors and the exhaustion of the technical resource of objects and control systems are common in present Russian circumstances. They urgently require guaranteed reliability and quality of control, including upgrades of project, operational, and modernization control capacities. One needs methods and technologies for the evaluation of control systems, and to ensure their optimality, functional and operational reliability, efficiency, fault tolerance, and survivability are necessary under the following conditions:

- The lack of priority information about the control object and external environment of its functioning, including in opposition conditions.
- There are many no stationarity factors to be challenging to take into account and their subjective character; x degradation (from failures, accidents) or necessity of targeted reconfiguration (revitalizing or developmental control). With the expansion of the functional loading, the control systems substantially become complicated. Many complexity factors appear in modern and advanced control systems.
- Multilevel controls, heterogeneity of description of subsystems by quantitative and qualitative models, different scales of processes in space and time, multimodality, multilink, decentralization and ramified nature, and general structural complexity of modern control systems and their control objects:
  - The presence of uncontrolled coordinate-parametrical, structural, regular, and singular impacts, including active counteraction in a conflict environment;
  - The use of deterministic and probabilistic models for the description of uncertainties of information about the vector of the state and parameters of the system, about properties of errors of measuring and environment; and
  - Non-linearity, distributed parameters, delay in control or object dynamics and sudden impacts, high dimension of models, and others.

3.9 THE LARGE-SIZED STRUCTURE OF CONTROL SCIENCE AND TECHNOLOGIES

Adaptive, robust, predictive, and other control methods developed in the theory of control are intended to consider the incompleteness of dynamics by obtaining the missing information during the training stage or in real time. The use of AI to expand the capacity of complex control systems by covering tasks with unknown or quantitative models is no longer valid for some moment of functioning, as well as studies where quantitative models are less efficient than the use of AI models (like in action planning tasks) or can be used in conjunction with AI models.

3.9.1 Neural Network Control

Finding minimum distortion of adversarial examples and thus certifying robustness in neural network classifiers is known to be a challenging problem. Nevertheless, recently, it is possible to give a non-trivial certified lower bound of minimum distortion, and some recent progress has been made toward this direction by exploiting the piece-wise linear nature of rectified linear unit (ReLU) activations. However, a generic robustness certification for exits {general} activation functions remains largely unexplored. To resolve this issue, a new system known as CROWN, a general framework to certify robustness of neural networks with general activation functions, is used. The novelty in this algorithm consists of bounding a given activation function with linear and quadratic functions, allowing it to tackle general activation functions including but not limited to the four popular choices: ReLU, tanh, sigmoid, and arctan. In addition, it is possible to search for a tighter certified lower bound by exit {adaptively} selecting appropriate surrogates for each neuron activation. Experimental results show that CROWN on ReLU networks can notably improve the certified lower bounds compared to the current state-of-the-art algorithm Fast-Lin while having comparable computational efficiency. Furthermore, CROWN also demonstrates its effectiveness and flexibility on networks with general activation functions, including tanh, sigmoid, and arctan. To the best of our knowledge, CROWN is the first framework to efficiently certify non-trivial robustness for general activation functions in neural networks.

Various AI means—neuro-net, evolutionary, logical, and others—can be used for action planning tasks and control in general. Each group has its advantages and disadvantages, especially about real-time requirements, and ensures the implementation of upper levels of heterogeneous control over complex systems. The intensive development of technical systems and processes (networking, miniaturization of sensors, controlling devices and calculators, improving their performance, etc.) puts new requirements for modern control systems. It opens new opportunities both at the level of embedded control systems of different scales and at the level of group interaction of decentralized multi-agent systems. Research and development are currently being done to transition from robots operating in an uncertain environment, but equipped with the operator interface (supervisory unmanned arial vehicle (UAV)) to intelligent robots. For that, one needs less expensive robots based on a modular principle of their construction and miniaturization, solving the problems of sensitizing, environment modeling, achieving the goals of the robots control team and extension of application scope. Even in agriculture and road building, radical transformation of standards requires robots with high-precision navigation and more intelligent control. Examples of critically important technological processes and smart control objects are the large-scale infrastructural systems of the power grid project/power industry.

Considering for a power grid project x is an inefficient structure of electro-network grids and generating capacity, x lack of energy saving in electricity consumption, x technological and commercial losses in electric networks, x a technological backwardness and high degree of wear of equipment a high level of monopolization of power markets, x vulnerability of electric power systems to terrorist and cybercrime threats, and others require developing the models of the complex infrastructural dynamic systems.

Program control u(t) Robust control (when parametric perturbations and uncertainties), adaptive, multimodal, and predictive control (when parametric and structural perturbations and uncertainties) Intelligent control (without goal setting) is described below:

Adaptive, multimodal, and predictive control (when parametric and structural perturbations and uncertainties) and Intelligent control (without goal setting):

in case of failure of subsystems or inadequate dynamics equations, reconfiguration, action planning, 3D-scenes analysis, learning Environment Intellectual control (with goal setting);

revision of the control quality goals and criteria, reflection and collective behavior, strategic interaction with other control systems) Position control u(t, x) (when coordinate perturbations) Automated control Man-machine level with the support using artificial intelligence Control object (a group of interacting objects) Regulation u(x). Heterogeneous control of complex systems Control based on logical-reactive (production) knowledge model in the so-called expert, recommender, or decision-making support systems which require to be enhanced with new features:

Co-operating with other means of control systems intellectualization (artificial neural networks, genetic algorithms) and algorithms of adaptive, robust, and predictive control, x reduction of interface complexity of logical control systems with the external physical world by combining methods of symbolic and multimedia presentation and knowledge processing;

Integration of quantitative and qualitative models with ontologies of different subject domains that characterize the problem situation.

Some advantages and disadvantages of AI means are explained in Table 3.1. There are different ways of combining different AI means. For example, one can integrate the neuro-reactive and logical-reactive (production). AI means with the first-order logical methods of intelligent control from 1,7. The latter methods can treat a wider stratum of knowledge, while the first two means support "reasonable" behavior based on providing the most superficial heuristic reactions of a control system for changes

TABLE 3.1
Some Changes to Consider before and after IO

Before IO	After IO
Single discipline	Multi-discipline
Serial process	Parallel process
Reactive	Proactive
Manual	Automated
Episodic	Continuous
Familiar technology	New technology
Single component	Integrated value chain

in an environment or the controlled object. Logical-reactive level (sometimes with its numerous "if-then" rules) especially needs verification of knowledge presentation. Productional rules of Boolean type with constructive semantics can reduce the verification of the knowledge base to the dynamic analysis of automata networks. This analysis is additionally simplified in the class of automata monotonous w.r.t. the state by applying for the mathematical models' properties transfer.

The critical problem in AI is the automatic estimation of the irrelevance of knowledge because a deficit information and a surplus information cause the degradation of intelligent control systems. Recent advances in intelligent control include the automation of searching for ways to achieve control goals given externally, while the automation of goal setting and revision of control quality criteria are insufficient. It is now recognized that improving only "machine components" in developed human–machine systems is not enough for the desired essential increase of their efficiency. The goal of creating anthropocentric systems can be achieved by directing the efforts of engineers and scientists on improving the intellectual component of the "system-core" in the anthropocentric approach as a built-in set of algorithms for embedded computers together with algorithms of operator activity, referred to as "onboard intelligence."

Correctness requires further consideration. First and foremost, the onboard intelligence is needed in aviation, especially in combat situations, typical for fighters, i.e., in the circumstances of the most aggressive external environment and tight timing constraints for the crew. Onboard intelligence is a functionally integral complex that aims to fulfill all aircraft tasks9. Scientific and technological advances in this field will also be helpful in other AI applications in the conditions of a multi-criticality, uncertainty, and risk to improve control quality in a situation of information overloading the operator, limited time, or stress. Development of practically useful onboard decision-making support expert systems, including those based on fuzzy logic and case-based reasoning by analogy, has reached the practical stage of building the models and prototypes. They are intensively developed to create the manned combat aircraft of the 4++ and 5th generations and combat UAVs. Their fragments already appear on the modernized fighters plane of the 4++ generation. In foreign developments, they are planned to be used onboard in the new USA fighters F-22, F-35, modernized F-16, F-15, F/A-18, and helicopters, which have several onboard intellectual systems of tactical decision-making. The research results, the improvement of onboard computers, cockpit displays board, controls, and other avionics control systems give the next-generation aircraft/helicopter constructors to design and realize a new type of-onboard computer systems. These systems will support tactical decision-making (the prompt appointment of the current purpose of flight and choice of a rational way of achieving the goal). Solving such tasks on past generations aircraft could only be completed by the individual efforts of the crew.

Further, some questions of intellectualization of automatic control systems in the form of fuzzy regulators and combining them with other AI means can be considered in detail. Note that the first regulators developed in Greece in the 3rd century BC partly can be regarded as the fuzzy controllers described linguistically with logical operations. Today, fuzzy control systems have many practical applications in transport, energy, oil and gas, metallurgy, medicine, and other industries and household appliances. It considers four basic types of regulators: logical-linguistic, analytical, learned, and proportional-integral differential fuzzy.

3.10 CHANGE IN DIGITAL OILFIELD MANAGEMENT (DOF)

DOF is seen as something unknown; it is even a threat to the strange and uncertainty. Some changes have been considered before and after the implementation of input/output (IO).

Today's digital oilfield has advanced, mainly in data storage capacity, standards acceptance and usage, computing power, and interconnectivity. Vendor solutions now use standard technologies such as Relational Databases for storage and common interfaces such as Open Platform Communications (OPC) and others. Because of these advances, new things have happened that have allowed us to interconnect systems.

3.11 LAYERS OF DIGITAL OILFIELD

The four layers of digital oilfield are hardware layer, operations layer, integration layer, and executive layer. These layers are visualized below; note that the middle/Integration layer is also called a field data capture system/layer (FDCS), and the executive layer is called the production data management system. It is narrated in flow logic how network data management server transfers e-FACS production data to executive management horizontally. Vertically, OPC server/e-FACS production data are sent to e-FACS field data capture server. Similarly, this OPC server horizontally forward these data to production account and reservoir data server. e-FACS operating system sends data horizontally to the field personnel/supervisory control and data acquisition (SCADA) system via scheduler and database. e-FACS operating system control field device, flow meters, computer command, programmable logic controller (PLC), and remote telemetry unit (RTU). This control logic is elaborated at network data management server as shown in Figure 3.1.

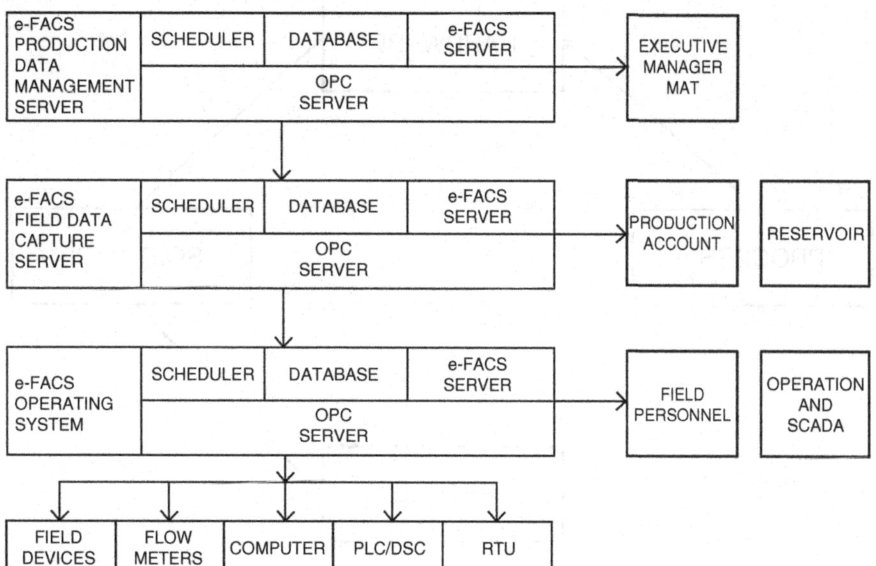

FIGURE 3.1 Network data management server.

3.12 PRODUCTION DATA MANAGEMENT SYSTEM

The hardware layer is the physical layer that monitors processes logic directly. The hardware layers include the following hardware:

- Subsea Xmas tree control systems
- Flow computers for measurement of flow
- Programmable logic controllers
- Distributed control systems
- Transmitters of all sorts of process operations

The hardware monitors the process of the DOF and is vital to recording what is occurring in the field, whether the area is onshore reading a tank level, offshore reading a down-hole pressure or temperature, or on the seafloor recording data from the well.

The operations layer is the first software layer where data are read from devices in the field and presented to users to determine what action to take. The most prevalent user at the operations level is the operator, who monitors the processes via a human–machine interface, typically letting them know if the methods are within tolerances and alert them if they are not so they can correct the issue.

Figure 3.2 explains the process of the DOF at every level, but the operations layer is the lowest level of software, which connects the hardware, the people, and processes together to make direct and actionable decisions. This is typically called the SCADA layer. However, in today's world, a SCADA system is required to do much more than just acquire data and perform supervisory control. At this level, resources

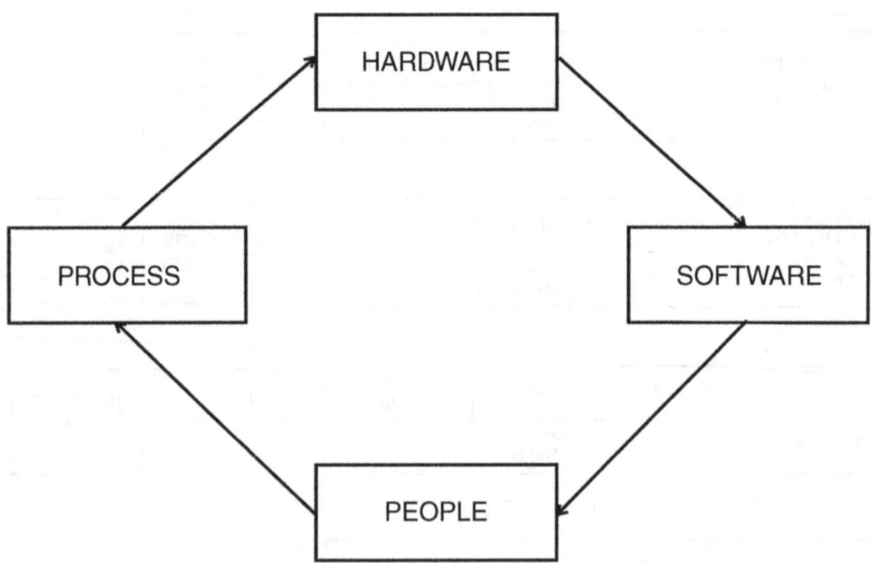

FIGURE 3.2 Digital oilfield execution process.

in roles such as field personnel, measurement techs, SCADA managers, and asset managers all use the system to view the information available in the network server. All of whom would like to see the data in possibly a different way. It is called the displaying of data in different ways of role-based displays/visualization.

Also, note that companies can have several, even hundreds of, operational centers where data are collected from hardware, recorded, and displayed in different ways. This had always executed in practice; in some cases, operational data are replicated to a central or regional server. The central or regional server is typically the FDCS/ Server. The FDCS may be located at the corporate site or a restricted site.

The FDCS usually gathers data from many different systems and integrates that data, not just from operational systems. The FDCS will get the following data from a system:

- Real-time history from data historians
- SCADA systems
- Measurement systems
- Production estimates plan
- Well testing data
- Third-party partner data
- Weather data

The term "digital oilfield" has been used to describe a wide variety of activities, and its definitions have encompassed an equally wide variety of tools, tasks, and disciplines. All of them attempt to explain various uses of advanced software and data analysis techniques to improve the profitability of oil and gas production operations.

The design of workflows and processes is a discipline that has historically been practiced by production planning engineers or operations management specialists. The language of oil and gas production has changed to reflect the adoption of their techniques. Traditional segmentation of responsibilities among functional lines (e.g., reservoir, drilling, well completion, and production engineers) is being integrated into engineering workflows and business processes that more broadly reflect the corporate objectives to be achieved (e.g., reservoir surveillance, well test validation, and production optimization). Digital oilfields, in one sense, comprise sets of workflows that allow fast, collaborative execution of interrelated tasks among distributed (virtual) teams, with the result that is optimal, efficient, and more profitable.

3.13 INTEGRATED ASSET MANAGEMENT

Integrated asset management (IAM) is a specialty alternative asset management company enabling investors to reduce risk and enhance returns. Founded in Toronto in 1998, Integrated Asset Management Corp. (IAM) is a public company (TSX: IAM) with the majority owned by the management.

Oil and Gas practice implements IAM solutions to provide visibility to assets and field equipment. It helps oilfield services companies and oil and gas distributors to maximize return on assets. Process harmonization improves operational efficiency while optimizing the utilization of assets and reducing operating and maintenance costs.

It will enhance the maturity of asset management processes for seamless collaboration across enterprise functions—including sales, design, and manufacturing. It uses predictive analytics and decision trees to identify potential failures and maintenance requirements. Our approach supports inventory management.

Another early benefit realized through integrated asset modeling was the ability to account for the uncertainty of model inputs by generating probability distributions for uncertain inputs and using Monte Carlo analysis to generate an equivalent distribution of outputs, along with their expected values. In general, whether deterministic or stochastic, the ability to automate the execution of multiple scenarios was immediately seen as a critical advantage of integrated asset modeling, which continues to influence the design of new workflows.

BIBLIOGRAPHY

Frederic, Biya Motto, Roger Tchuidjan, Benoît Ndzana, and Jacques Atangana. Model of Reactive Electromotor with Full Phase Functioning Regime, *IOSR Journal of Applied Physics*. 2019. 11 (6) (Ser. II): 7–12.

Integrated Asset Management Corp. "Leadership in Alternative Integrated Asset Management. Annual Report." Mink Capital, Canada. 2016.

Kumar, R. *Artificial Intelligence with Python* (Python Technology Book 1). Independently Published. 2020. ASIN: B0874Q5N65.

Petro-Wiki. n.d. "Digital Oilfields." Published by spe.org, USA. Future trend Para-2. Last updated. 2011 February 18. https://petrowiki.spe.org/Digital_oilfields.

Vámos, T. Artificial Intelligence, Automatic Control, and Development, *Hungarian Academy of Sciences*. 1980. 13 (11): 8–11.

Vasilyevich, S. N., A. Yu. Kelina, Y. I. Kudinov, and Fedor Fedorovich Pashchenko. Intelligent Control Systems, *Procedia Computer Science*. 2017. 103: 623–628.

4 The Components of Trusted Artificial Intelligence

4.1 INTRODUCTION

One of the main difficulties with analyzing the ethical impact of artificial intelligence (AI) is overcoming the tendency to anthropomorphize it. The media is enthralled by images of machines that can do what we can, and often, far better. People are bombarded with novels, movies, and television shows depicting sentient robots, so it is not surprising that we associate, categorize, and define these machines in human terms. While people associate human activities and abilities to devices, it becomes problematic when this anthropomorphization is attached to human moral activities, such as trust. AI makes a computer, a computer-controlled robot, or software think intelligently, similar to an intelligent human being.

AI systems can identify by studying how the human brain thinks and how humans learn, decide, and work while trying to solve any problem and then using the outcomes of the study to develop intelligent software and systems.

AI systems are nowadays increasingly being used to support the human decision-making process. At the same time, AI promises to deliver valuable insights and knowledge across many applications; the broad adoption of AI systems will heavily rely on the ability to trust their output. Humans trust in technology based on their understanding of how it works and the assessment of its safety and reliability. To modify a decision made by an algorithm, one needs to know that it is reliable and fair, can be accounted for, and will cause no harm.

AI has centered mainly on the following components of intelligence: learning, which is classified in to several forms; reasoning; problem-solving; perception; and language understanding.

4.1.1 Trusted AI

Trustworthy AI governed the idea that trust builds the foundation of societies and sustainable development. Therefore, individuals, organizations, and institutions will only ever be able to realize the full potential of AI if they can establish trust in its development, deployment, and use.

Through the open source (sponsored by IBM), they bring trust back into AI. Increasingly, dimensions of faith, including fairness, robustness, and explainability, are important metrics that help evaluate AI model behavior.

DOI: 10.1201/9781003307723-4

There are four types of AI in the industry sector: reactive machines, limited memory, theory of mind, and self-awareness.

4.1.2 Application of AI Systems

AI is the simulation of human intelligence designed by machines, especially computer-based systems. These processes consider learning, reasoning, and self-correction. Other applications of AI include expert systems, machine vision, and speech recognition.

Since AI is about synthetically achieving intelligence to replicate human nature, behavior, and capabilities and make the machines operate as if operated using human intelligence, thus making devices more "human like," it becomes imperative that AI should have the following.

4.2 COGNITIVE ABILITIES

If AI wants the machine to interact like humans, it will have to think and work like humans. Cognitive computing is commonly used to describe AI systems that simulate human belief. Several AI technologies are required for a computer system to build human mental models that mimic human thought processes, including machine learning (ML), deep learning, neural networks, natural language processing (NLP), and sentiment analysis.

- The computer-generated intelligent machine is equipped with NLP for successful interactions.
- Automated reasoning for analyzing the database is stored in the system.
- ML will feed the machine with present inputs, and after training sessions, the machine will be able to speak and learn language owing to the pre-fed data to respond more accurately and less like a robot.
- Rationality and introspection developed with more training.

4.2.1 Database

The machines may be made more intelligent and introspective rationales only by incorporating and supporting them with proper databases and information, and feeding them in the form of rules to process information and help find solutions for humans.

4.2.2 Hardware

Graphic Processing Units, aka GPUs, are essential for AI hardware to extend from one approach to a physical entity using the latest technology.

4.2.3 Framework

For better ML processing, a sound framework is needed. Python software, R, and Azure Machine Learning Studio are the most commonly used.

4.2.4 Role of APIs for AI Services

The role of text classification, image classification, sentiment analysis, etc., can be achieved using application programming interface (API) services.

4.2.4.1 AI Applications for User's End

While a lot facilitates AI in creating an exciting and successful outcome, the prerequisites are AI components. At Sphinx Worldbiz, the AI experts generally dedicatedly use three best-of-class infrastructures toward successful AI results in bringing the technology closer to its end users.

4.3 PHILOSOPHY OF AI

To use the power of computer systems, the curiosity of humans leads them to wonder whether it is possible that *a machine thinks and behaves like a human brain.*

Thus, the development of AI started to create similar intelligence in machines that we find and regard high in humans.

4.3.1 Goals of Any AI System

To Create Expert Systems: The systems that project intelligent behavior learn, demonstrate, explain, and advise their users.

AI Human Intelligence in Machines: Creating new subsystems that understand, think, learn, and behave like humans.

4.3.2 Contribution of AI Systems

AI is a field of science and technology based on computer science, biology, psychology, linguistics, mathematics, and engineering. A major thrust of AI is in developing computer functions associated with human intelligence, such as reasoning, learning, and problem-solving.

One or multiple areas that can build an intelligent system are represented by a flow diagram of an AI system in Figure 4.1.

4.3.3 Programming with and without AI

The purpose of programming without and with AI is different in the following ways:

AI programs can resolve new modifications by putting highly independent pieces of information together. Hence, it can modify even a minute part of the knowledge of the program without affecting its structure. Modification is not quick and straightforward. It may lead to affecting the program differently, quick and easy program modification may not be always desirable.

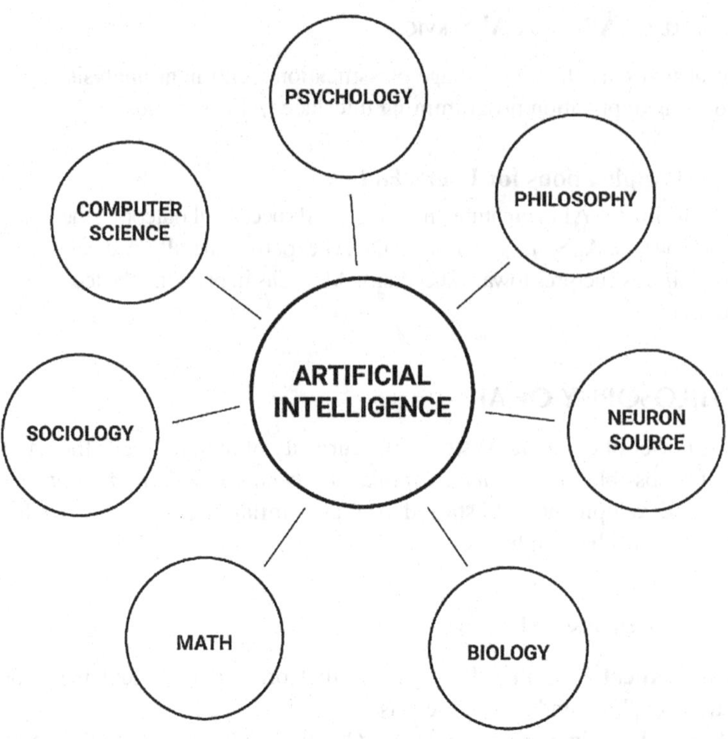

FIGURE 4.1 Flow diagram of an AI system.

Programming without AI	Programming with AI
A computer program without AI can answer the specific questions it is meant to solve.	A computer program with AI can answer the generic questions it is meant to solve.
Modification in the program leads to change in its structure.	AI programs can absorb new modifications by putting highly independent pieces of information together. Hence, it can modify even a minute piece of information of program without affecting its structure.
Modification is not quick and easy. It may lead to affecting the program adversely.	Quick and easy program modification.

4.3.4 Building Four Critical Components of a Trusted AI System

Building trust in AI systems is hard. The AI systems are explained under a four-component concept. The four pillars of Trusted AI are:

1. Fairness
2. Explainability
3. Robustness
4. Lineage

However, other components involved in building AI systems are as follows:

- Knowledge reasoning
- Planning
- ML
- Neural language processing
- Computer vision
- Robotics
- Artificial general intelligence

Trust and transparency are at the vanguard of AI conversation. While it understands the idea of trusting AI agents, it always points out the specific process to translate transparency and trust into programmatic constructs. After all, faith in AI means in the context of an artificial general intelligence system.

Trust is a foundational building block of humans in socio-economic dynamics as explained in Table 4.1.

During the last few decades, researchers have steadily developed a software mechanism for asserting trust on specific applications. When robots ultimately drive a plane that flies on auto-pilot mode, we intrinsically express confidence on the creators of a particular software application. Their behavior is uniquely determined by the code workflow, which is inherently predictable. The non-deterministic core of AI systems breaks the pattern of conventional software applications and introduces new dimensions to authorize trust in AI agents.

Trust is a dynamic process derived from the process of reducing the risk. In the software development stage, trust is built through mechanisms such as suitability, testability, documentation, and many other elements that help to establish the reputation of a part of the software. In conventional software applications, their functioning is altered by clear-cut rules indicating AI in the code; in the case of AI agents, their behavior is based on the knowledge that evolves. The former approach is deterministic and predictable, and the latter is non-deterministic and challenging to understand.

AI will be a relevant part of future ML systems; it is essential to establish trust in AI systems. The idea of trust in AI systems remains favorably prejudiced and hasn't been incorporated as part of ML frameworks or platforms. The way is to measure the AI trust. Trusted AI is explained in detail in the following subsection.

TABLE 4.1
Foundational Building Block of Humans in Socio-Economic System

Ethics of Algorithms	Ethics of Data	Ethics of Practices
Respect for human autonomy	Human-centered	Responsibility
Preparation of harm	Individual data control	Liability
Fairness	Transparency	Code of regulation
Explicability	Accountability	
	Equality	

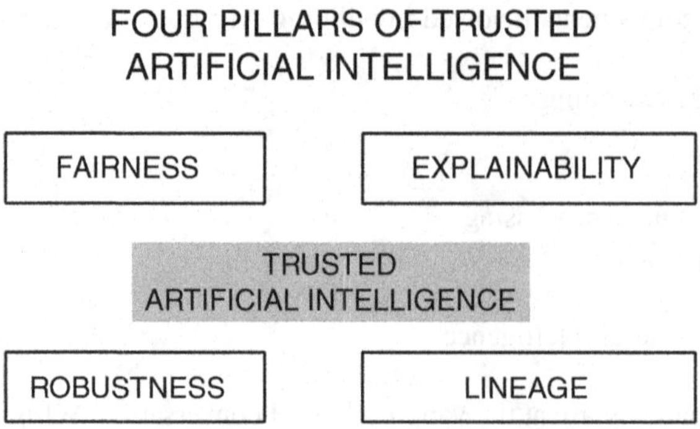

FOUR PILLARS OF TRUSTED ARTIFICIAL INTELLIGENCE

FAIRNESS		EXPLAINABILITY
	TRUSTED ARTIFICIAL INTELLIGENCE	
ROBUSTNESS		LINEAGE

FIGURE 4.2 A diagram showing fundamental pillars of trusted AI.

4.3.5 The Four Pillars of Trusted AI

Trust in human interaction is based not only on the better interpretation of specific actions but also on social knowledge built throughout centuries. It understands that behavior is discriminatory not only by judging it in realtime but also by factoring in a socially accepted concept that discrimination is derogatory to human beings. These ideas need to be extrapolated to the world of AI. A team of IBM engineers proposed four fundamental pillars to trusted AI which are shown in Figure 4.2.

4.3.6 Fairness

The fairness of AI is typically associated with the minimization of bias in AI agents. Preference can be described as the mismatch between the training data distribution and a desired fair distribution. Unwanted bias in training data can result in erroneous results. Establishing tests for curating, identifying, and minimizing bias in training datasets should be pivotal to establishing fairness in AI systems. Righteousness is more relevant in AI apps with a tangible social impact, such as credit or legal applications.

4.3.7 Explainability

To arrive at specific decisions by AI is another fundamental principle of trusted AI. In order to understand meaningful explanations about the knowledge of AI models Which reduces uncertainty and helps to quantify their accuracy. While explainability might be seen as a prominent factor to improve the trust in AI systems, its implementation is far from trivial. There is a natural tradeoff between the accuracy of AI models and the explainability. Highly interpretable AI models tend to be very simple and, therefore, not on the condition that it is accurate. From that outlook, establishing the right balance between accuracy and explainability is essential to improve the trust in an AI model.

4.3.8 Robustness

The concept of AI robustness is determined by two underlying factors: safety and security, which are elaborated below:

4.3.8.1 Safety

An AI system might be explainable and fair but still risky to use. AI safety is customarily associated with the ability of an AI model to build knowledge that incorporates policies, societal norms, or regulations that correspond to well-established safe behaviors. More and more safety of AI models is another crucial element of trusted AI systems.

4.3.8.2 Security

AI models are highly vulnerable to all types of attacks, including most of them based on antipathic AI methods. The exactness of AI models is directly correlated with their quality of being easily attacked by a minor disturbance of motion on the input dataset. That relationship is often exploited by a person not responsible for the action that can try to alter specific datasets to change/influence the behavior of AI models. The key to establishing trust in AI systems is testing and benchmarking AI models against adversarial attacks. Various computer companies have been doing some exciting work in this area.

4.3.9 Lineage

AI models are constantly developing, making it challenging to locate their history. Setting up and tracking the provenance of training datasets, hyperparameter configurations, and other metadata artifacts over time are essential to select the lineage of an AI model. Understanding the origin of AI models helps us establish trust from a historical perspective that is different to achieve by factoring in fairness, explainability, and robustness alone.

4.4 A FACTSHEET FOR AI SYSTEMS

The subject of transparency and disclosures in AI systems is a very nascent area but it is the key to the mainstream adoption of AI. When information sheets for hardware appliances or nutrition labels in foods are used, it is necessary to establish a factsheet for AI models. In the IBM system, it is essential to have a Supplier's Declaration of Conformity (SDoC, or factsheet, for short) that helps provide information about the four critical pillars of trusted AI. SDoC methodology should help answer basic questions about AI models such as the following:

- The dataset used to train the service should have a datasheet or data statement ready.

- The dataset and model should be checked for biases; they should describe bias policies, bias checking methods, and results. The mitigation method performed on the dataset to confirm is known as bias mitigation. Explainability of the algorithm outputs should be achieved (e.g., directly explainable algorithm, local explainability, explanations via examples).

4.5 TESTING METHODOLOGY

AI is a new technical discipline that develops theories, methods, technologies, and application systems for simulating the extension and expansion of human intelligence. The goal of AI directs and allows machines perform some complex tasks that require intelligent humans to complete. The device is expected to replace us to solve some complicated tasks, not just repetitive mechanical activity but some that need human wisdom to participate in it.

4.6 KEY TECHNOLOGIES OF AI

The key technologies of AI can be divided into the algorithm layer and the infrastructure layer from bottom to top. In the megastructure layer, there is basic hardware, including CPUs, GPUs, dedicated chips, and high-speed networks. On top of this basic hardware, algorithm frameworks can be created, such as Tensor Flow, Caffe, Mxnet, Torch, Keras, PyTorch, and Theano. This organizational layer is the algorithm layer. The most well-known algorithm layer is ML algorithm, including a series of ML algorithms such as deep learning, transfer learning, general adversarial network, and reinforcement learning.

4.7 INFRASTRUCTURE LAYER

4.7.1 Basic Hardware

The basic hardware of AI infrastructure is divided into four categories: CPU, GPU, special-purpose chip, and high-speed network.

The following are the two similar cases: CPU and GPU. CPU is mainly optimized for serially executed tasks, while GPU improves complex graphics and image calculation algorithms. The difference is that the CPU executes serially, and GPU is a smaller, more efficient computing unit that processes calculations in parallel. Furthermore, a special chip is specially developed and designed for AI algorithms, known as Google's TPU chip.

To fully utilize the ability of AI, a high-speed network is needed. During the training and calculation of some complex data models, huge network bandwidth guarantees are also needed. Presently, the network has become an essential part of ML performance. Now, it commonly had 10G, 20G, and 40G networks. With the advent of Infiniband network technology, the network will provide a broader and faster channel for AI learning and training in the future.

4.7.2 Framework of Algorithm

The second layer in the infrastructure is the algorithm framework. The algorithm framework can be understood as the framework for running algorithms, just like a building framework on which it can run the business. Google developed a very famous framework called Tensor Flow, which is friendly, fast, and convenient. Tensor Flow lets us implement related AI algorithms and run these methods.

4.7.3 Algorithm Layer

First of all, ML is a core concept in AI. All these must be learned, and human knowledge transfer is also carried out through learning. It can realize ancestors' knowledge and then create new knowledge by inference. It can also be hoped that the machine has such ability: By learning the previous information, it is more like having intelligence and can react accordingly to new input in the future, which is called machine learning.

4.8 COMPONENT ANALYSIS

In the component analyzing phase, the first set is generated, which contains appropriate elements for each function and the corresponding attributes of each component.

4.8.1 Extracting Criteria and Weighting Criteria

In this stage, some criteria considering the dark sides of AI are taken out, and then the analytic hierarchy process (AHP) mechanism is routinely calculated as the weights of the components. This phase is divided into two sub-phases as follows.

4.8.1.1 Weighting Criteria

It is a tool to compare the proposed alternatives quantitatively. This means ranking which criterion is most important for the decision-making process.

4.8.1.2 Extracting Criteria

Although AI provides notable changes and improvements for a network of digital businesses and facilitates innovative services and digital transformation, AI's plentiful dark sides present a tremendous risk for individuals, organizations, and society. The initial and most crucial step is identifying and classifying such criteria to address these dark sides. So, a list of potential AI dark sides is provided as follows:

- Energy consumption
- Data issues
- Security and trust
- Privacy
- Fairness
- Safety

- Beneficial
- Predictability
- Explainable AI
- Complexity issue
- Monopoly
- Responsibility challenges

The above-mentioned are explained in more detail in the following subsections.

4.8.1.2.1 Energy Consumption

One of the darker sides of most ML algorithms is high energy consumption. These days, most ML algorithms rely on iterative policies instead of fixed policies, which leads to increased energy consumption and energy wastage. Moreover, this problem is increased by a growing number of learning models requiring more learning motive iterations. For example, deep learning methods require GPUs' high computational natural language power compared to other methods. Contamination of the environment and global warming are other side effects of increased computational power usage.

4.8.1.2.2 Data Issues

This type of AI invests in data-driven algorithms to construct ML models. There may be many problems in data that lead to many difficulties in data-driven ML in many situations. Some of these problems are as follows.

Big Data

The magnitude of datasets collected in a wide range of systems such as Internet of Things (IoT) and augmented reality (AR) is increasing. Defining ML algorithms that can operate online leads to a challenging problem with massive data. It can use a wide range of methods, such as sampling, distributed processing, and parameter estimation, to obtain required information from data.

Data Incompleteness

Deficient data is a challenging problem in ML algorithms. Preliminary data in every dataset may misdirect the algorithm to learn unsuitable models. This challenge develops uncertainties during data analysis if incomplete data are not considered during the data analysis step. Many implication methods exist for this problem. An initial viewpoint is to fill a training set with the most often observed values or build learning models to foretell possible values for each data field based on the experimental values of a given instance.

Other issues in this realm are data heterogeneity, insufficiency, uncertainty, originality, inaccuracy, imbalanced data, data dynamicity, and high-dimensional data.

4.8.1.2.3 Security and Trust

Security and trust are two critical issues that have received much attention in recent years. In IT, these issues have two dimensions as follows. Most of the studies only focused on employing ISs to design fixed systems. It is where something that every piece of software may be hacked or cracked. AI experts do not consider these issues because the development of ISs in critical systems is in the early stages. For example, the trusted data in data-driven ML will construct a model based on it. When these data are untrusted, the ML model is also untrusted. A spiteful person may swap trusted data with untrusted data during the software lifecycle, and this phenomenon may occur in every data-driven approach. An emerging field called adversarial ML was the first attempt to solve some security problems in data-driven ML. It may consider attack mechanisms to manipulate the evolutionary processes for other AI-based methods such as genetic algorithms.

4.8.1.2.4 Privacy

The privacy topic in AI has different dimensions. Many ISs have been developed based on extensive data analyses, data sciences, and data-driven methods in recent years. These procedures are fed by the data of a considerable number of consumers. During the implementation of these methods, three different roles are possible, as explained below:

- A role for data manifestation and data owners (or contributors)
- A role for data analyzers and model skillful person
- A role for result visualizers

Generally, a programmer has all those roles during the designing of ISs. However, different entities may play the mentioned roles in industrial projects, and these entities may not be trusted considering privacy issues. Federated learning is one of these efforts. Network providers agreeing upon standards of operation learning is a ML technique that trains an algorithm across multiple decentralized computers without swapping data among them, thereby addressing reproving topics such as data privacy, data security, and data access rights.

4.8.1.2.5 Fairness

In AI, a given algorithm is fair if its results are independent of sexual orientation, gender, and ethnicity. The logic behind this issue is that many people have disabilities, and gender must not add rights to users. This issue becomes more challenging when some attributes such as gender and race are sensitive in the culture of humans.

AI is one of the successful algorithms in intelligent managing systems. In mission-critical, real-world environments, there is minor tolerance for failure that can cause damaging effects on devices and humans. In the real-world environments, current tasks, the two main methods mentioned in the literature, are not sufficient to support the safety of humans. The first approach is to control the output of an information system (IS) while considering the safety of humans. In the second approach proposed, computation in the internal parts of an intelligent agent will manipulate using some weights after careful consideration of the safety of humans.

4.8.1.2.6 Beneficial

Soon, ISs will make better judgments than humans in many domains, including computer vision. An IS can analyze many diseases using image processing techniques that are more diligent than human vision. For more than one decade, humans have been trained to reach out to the works of their forefathers in simple expertise such as piping and constructing a town. This can quickly execute these jobs, which will control many things soon. In some situations, an IS may determine to do an action harmful to humans. Most existing systems with deterministic decision-making mechanisms may efficiently execute unhealthy decisions without considering human preferences. In these situations, beneficial AI computation can be applied. With this concept, a system is designated to behave so that humans are pleased with the outcome. In these systems, the agent is uncertain about humans' preferences and uses human behavior to extract information about human preferences.

4.8.1.2.7 Predictability

One of the most important issues in designing ISs is predictability. This issue becomes more when many management algorithms in different fields utilize ISs. In developing ISs, many factors exist which destroy the certainty ability. Some of these factors are explained here. Paradox and equivocation are two factors that exist in text, image, and voice, and therefore, a system with these types of input data cannot present a foreseeable result. Some concepts in computer sciences, such as Turing decidability, the Gödel theorem, and the strange loop theorem, prove unpredictable behavior in most systems.

4.8.1.2.8 Explainable AI

Explainable AI refers to methods that help human experts to understand the results of the solution they obtained. Many AI-based systems, known as best problem-solvers, such as deep learning, only focus on a mathematically specified goal system determined by the designers. Therefore, a human agent may not understand the system's output. This problem can be very demanding in military services and the healthcare system because a human agent should evaluate the rationale behind a decision. In the literature, there are some attempts to solve this problem.

4.8.1.2.9 Complexity

The complexity of ISs is increasing day by day. The previous versions of ISs invest in a limited set of solutions to execute their jobs, and therefore, their complexity is restricted to simple algorithms. Nowadays, the existing ISs use numerous learning algorithms. The complication related to the size and format of data was discussed in an earlier section.

In addition to the mentioned above, many challenges are associated with the complexity of input, computation, memory, output, etc. Novel approaches to solving complex problems in complex systems rely on digital twin technologies.

4.8.1.2.10 Monopoly

Many AI-based solutions require substantial computational power. A few software companies (IBM, Amazon, and Microsoft) and countries infuse AI and high computational power devices. For example, a few countries explore quantum computation, enabling artificial general intelligence and super-intelligence. This ability may lead to the appearance of a monopoly in the scope of AI. Those companies that can execute many AI-based algorithms can do valuable activities such as developing new drugs and treatments for diseases.

4.8.1.2.11 Responsibility Challenges

AI-based systems, such as self-driving cars, will act autonomously in this world. It will make superior decisions in many fields than humans in due course. A demanding question in these systems, such as self-driving cars, is: who is likely if a self-driving car is involved in an accident? This problem has many dimensions. It appears that many laws must be defined considering those ISs engaged in decision-making processes. From an algorithmic perspective, frameworks will need to extract the responsibility of all entities that are involved in decision-making processes—some interesting points related to responsibility issues covered for a specific case study.

4.9 FINDING COMPONENTS

A learning automata-based solution for the component selection in this phase, a learning automata-based algorithm, is utilized to find the knapsack problem formulated in the previous step. This algorithm gets iteration number, reward (a), and penalty (b) parameters and then searches with a solution space-based probabilistic search mechanism to find an appropriate solution. In writing this type of algorithm, the problem is modeled as a complete graph where each graph node corresponds to a component in the knapsack problem. Every node of the chart is available with a learning automaton with two actions of selecting either item to be placed in a knapsack or not. In this mechanism, agents are defined to activate learning automata. In each iteration of the algorithm, there are a few agents, each of which creates a solution. Initially, an agent is randomly placed on one of the graph nodes and activates the learning automaton of that node. Whenever a learning automaton is activated, it selects one of its two actions according to the probability vector of its actions. Afterward, the solution set is constructed based on the activity chosen by the learning automaton. Finally, the solution set is modified considering some criteria. This process is repeated based on the predefined iteration number.

BIBLIOGRAPHY

Li, Cheng-Hong, Rebecca Collins, Sampada Sonalkar, and Luca P. Carloni. "Design, Implementation, and Validation of a New Class of Interface Circuits for Latency-Insensitive Design." *2007 5th IEEE/ACM International Conference on Formal Methods and Models for Codesign (MEMOCODE 2007)*, 2007. 13–22. IEEE.

Peculis, Ricardo, Farid Shirvani, and Pascal Perez. "Assessing Infrastructure Systems of System Integrity." *22nd International Congress on Modelling and Simulation*, Hobart, 2017 December. 3–8.

Qureshi, Kashif Naseer, Gwanggil Jeon, and Piccialli, Francesco. Anomaly Detection and Trust Authority in Artificial Intelligence and Cloud Computing, *Computer Networks*. 2021 January 15. 184: 107647.

Tangi, Rajesh. *Cognitive AI...Simplified*, LinkedIn, New York. 2020.

Yeung, Joshua. "Overview of the Key Technology of Artificial Intelligence." Towards Data Science, Ontario, Canada. 2020 April 11. https://towardsdatascience.com/overview-of-the-key-technologies-of-artificial-intelligence-1765745cee3.

5 Industrial Internet of Things (IIoT)

5.1 INTRODUCTION

Industrial Internet of Things (IIoT) is a subject where all hardware pieces work together through connectivity to help increase manufacturing and industrial processes. IIoT significantly impacts established manufacturing companies' business models (BM) within several industries. Thus, it is being attempted to analyze the influence of the IIoT on these BMs, especially with respect to differences and similarities dependent on different industrial sectors:

- The manufacturing industry is facing changing workforce qualifications.
- The electrical engineering and information technology and communication technology agencies are mainly concerned with the importance of novel critical partner networks.
- Automotive suppliers predominantly exploit IIoT-inherent benefits in terms of increase of cost-efficiency.

5.2 BACKGROUND OF IIOT

Internet of Things (IoT) elaborates the extension of Internet connectivity between physical devices and everyday objects. Consist with electronics, Internet connectivity, and other forms of hardware such as sensors, these devices can exchange information and interact over the Internet and be remotely controlled and monitored. IoT sensors and devices can provide real-time data on machinery, piping, storage, transportation, and employee safety within the oil and gas industry.

The applications of IoT in the offshore oil and gas industry, as with other digital technologies, center on improvements in efficiency and operational safety. For example, by integrating the IoT into offshore platform equipment operation, employees can track and monitor equipment lifespan and other elements that may affect oil and gas production, such as various marine parameters like wave heights, temperature, and humidity. By applying this information, companies can effectively maintain an offshore platform through predictive maintenance, helping to detect equipment breakdowns before they occur. It, thus, leads to enhanced productivity with less downtime. In addition, it has the advantage of reducing the necessity for physical presence and inspection in the remote offshore complex. This type of application can be integrated to other digital technologies such as automation, artificial intelligence (AI), and big data analytics. Hence, it is an excellent example of how new technologies can positively impact offshore equipment availability.

DOI: 10.1201/9781003307723-5

If applied into a suitable data analytics system, the ability to collect real-time data via IoT can have powerful advantage of improving equipment efficiency, with only minor efficiency improvements resulting in notable increases in oil production. Oil production data captured in real time through embedded sensors linked to automated data transmission systems can enable oil companies to collect information from assets anywhere and make informed operational decisions. For example, companies may adopt real-time drilling strategies and decisions based on comparisons of real-time downhole drilling data and oil and gas production data from nearby wells. This data collection and integration type can improve production by 6%–8%.

The concept of IIoT first became popular in 1999 through the Auto-ID Center at MIT and related market analysis publications. Mr. Kevin Ashton observed radio-frequency identification (RFID) as one of the prerequisites for the IoT then. If all objects and people in the daily life cycle with identifiers, computers could manage and inventory controls them. Besides RFID, the tagging of things may be achieved through such technologies as near-field communication, barcodes, QR codes, and digital watermarking.

One of the first outcomes of implementing the IIoT would be to create instant and ceaseless inventory control. One more benefit of implementing an IIoT system is building a digital twin of the system. Utilizing this digital twin allows for further system optimization by allowing for experimentation with new data from the cloud base system without stopping production or sacrificing safety. It can refine the new processes until they are ready to be applied. A digital twin can also be a training ground for new employees who won't worry about real impacts on the daily live system.

Digital twins combine intelligence and data that represent the context, structure, and behavior of a physical system of any type, offering a shared boundary that allows one to understand past and present operations and make predictions correctly.

These are compelling digital objects that can optimize the physical world, significantly improving operational performance and business processes.

A complex system generates vast amounts of data and connects a group of systems through the Internet, and the data increase considerably. All of the data coming off of these devices are illustrative. That is, the data pinpoint the reason why it happens and when it happened. Data analytics increases the data to be predictive and communicate when something will happen, a failure, for instance. But data analytics doesn't predict how to improve the product to avoid further failure. However, any digital twin can predict a 3-D digital model of a bodily system that can do this.

The first industrial revolution was characterized by mechanical production powered by water and steam. The second revolution used a large labor force and electrical energy, while the third featured electronic and automated production. The term "industrial revolution" was first proposed in 2011 to describe the development of the German economy. This revolution is distinguished by its reliance on cyber-physical systems (CPSs) that enable communication and autonomous, decentralized decision-making. Industry version 4.0 is a collective term of technology and value chain organization concepts within the modular structured Industry 4.0. Some industry treats it as synonymous with IIoT.

CPS monitors physical processes, creates a virtual copy of the physical world, and makes decentralized decisions. Over the IoT, CPS communicates and cooperates with humans in realtime, and via the Internet of Services, internal and cross-organizational services are extended and utilized by value chain contributors.

5.3 EVOLUTION OF IIOT

The evolution of IIoT in the industry is described chronologically.

5.3.1 Industry 1.0 (1784)

The invention of steam engines kick-started Industry 1.0. However, the manufacturing was purely labor-oriented and tiresome.

5.3.2 Industry 2.0 (1870)

Assembly line production was the major breakthrough of the Industry 2.0. This invention was a significant relief for the workers as it minimized their labor to a maximum extent. Henry Ford, the father of large volume production and the assembly line, introduced Ford's car manufacturing plant to improve productivity using the conveyor belt mechanism.

5.3.3 Industry 3.0 (1969)

This period saw the advancement of electronic technology and industrial robotics, miniaturization of the circuit boards through programmable logic controllers, and industrial robotics to simplify, automate, and increase production capacity. However, the operations remained isolated from the entire enterprise.

5.3.4 Industry 4.0 (2010)

Introduction of Industry 4.0 fulfilled the vision of connected enterprise by interconnecting industrial assets through the Internet. The intelligent devices convey with each other and create valuable insights. IIoT brought the advantages of asset optimization, production integration, intelligent monitoring, remote diagnosis, thoughtful decision-making, and most importantly, the feature of predictive maintenance.

5.4 MAIN ARCHITECTURE OF IIOT

There is not a single consensus on architecture for IoT, which is agreed universally. Several companies have proposed different architectures.

5.4.1 Three- and Five-Layer Architectures

The primary architecture is a three-layer architecture [3–5], as shown in Table 5.1. It was initiated in the early stages of research in this domain. It has three layers: network, perception, and application layers.

TABLE 5.1
Architecture of IoT

A	B
Application Layer	Business Layer Application Layer
Network Layer	Process Layer Transport Layer
Perception Layer	Perception Layer

5.4.1.1 Perception

The *standard perception layer* is the physical layer, with sensors for sensing and gathering information about the environment. It smells some physical parameters or identifies other intelligent objects in the background.

5.4.1.2 Network

The *network layer* is accountable for connecting to other network devices, bright things, and servers. Its characteristics are also used for transmitting and processing sensor data.

5.4.1.3 Application

The *application layer* is accountable for delivering application-specific services to the user. It defines various petitions in which the IoTs can be located, for example, smart homes, smart cities, and innovative health. The components of IoT architecture (A: three layers, B: five layers) are shown in Table 5.1.

The three-layer architecture can define the main idea of IoT, but it may not be sufficient for IoT because research often focuses on more refined aspects of IoT. Many more layered architectures are also reported in the literature. One is the five-layer architecture, which additionally considers the processing and business layers. The five layers are perception, processing, transport, application, and business layers (see Table 5.1). The role of the acumen and application layers is the same as that of the architecture with three layers. The functions of the remaining three layers are outlined as follows:

- The *transport layer* always transfers the sensor data from the perception layer to the processing layer and vice versa through other networks such as wireless, 3G, LAN, Bluetooth, RFID, and NFC.
- The *processing layer* is known as the middleware layer. This layer stores, analyzes, and processes a considerable amount of data, which is received from the transport layer. It can manage and provide varied services to the lower layers. It employs many technologies such as cloud computing, databases, and extensive data processing modules.
- The *business layer* controls the whole IoT system, including application, business, and profit models, and users' privacy.

- Another architecture proposed by Mr. Ning and Wang is encouraged by the layers of processing in the human brain. It is inspired by the ability of human beings to think, feel, remember, make suitable decisions, and react to the physical environment. This architecture had three parts. The first part is the human brain, analogous to the processing and data management unit or the data center. The second is the spinal cord, similar to the distributed network of data processing nodes and intelligent gateways. The third is the network of nerves, which corresponds to the networking components and sensors.

5.4.2 Cloud- and Fog-Based Architectures

Edge—Network—Cloud Requirements for an IIoT architecture are scalability, real-time capability, interoperability, and data protection and security. Central means sensors, actuators, and intelligent devices collect data (edge computing) and send it to servers (network). Intelligent algorithms are further processed into action-relevant "smart data" at the cloud computing level. They then develop the basis for automated processes. Major cloud contributors such as Microsoft, Amazon Web Services, and Google now create IoT platforms that smooth the development and administration of IIoT applications. Solving such a problem is ideal for entry-level people in the industry.

The flow layout of IIoT consists of several interfaces for connections to other devices. **IoT** device has an I/O interface for sensors and, similarly, for Internet connectivity, storage, and audio/video. **IoT** devices collect data from onboard or attached sensors and sense data communicated to other devices or cloud-based servers, as explained in Figure 5.1.

The solution environment, the executed use cases, and the technologies utilized in the IIoT stack are still different. Therefore, users' success increasingly requires seamless integrations, other technology portfolios, and expertise in the field.

IIoT refers to utilizing IoT technology in industrial production. In the German-speaking world, the IIoT is mentioned as Industry 4.0.

Disparity to the IoT in the private domain, the focus in the industrial sector is on the networking of machines. The fundamental concept of the IIoT is thus to integrate machine learning and big data technologies and thus considerably increase the success of companies. However, complexity and technical requirements are much higher in IIoT than in IoT (Figure 5.2).

The IIoT gives power through cybersecurity, edge computing, cloud computing, mobile technologies, machine-to-machine, 3D printing, advanced robotics, big data, the IIoT, RFID technology, and cognitive computing. Five of the most important ones of IIoT are described in the following subsections.

5.4.2.1 Cyber-Physical Systems

There are many definitions for CPS. CPS is a system comprising interacting physical and digital components. The system may be managed in one place or distributed, providing a combination of sensing, computation, and control networking functions

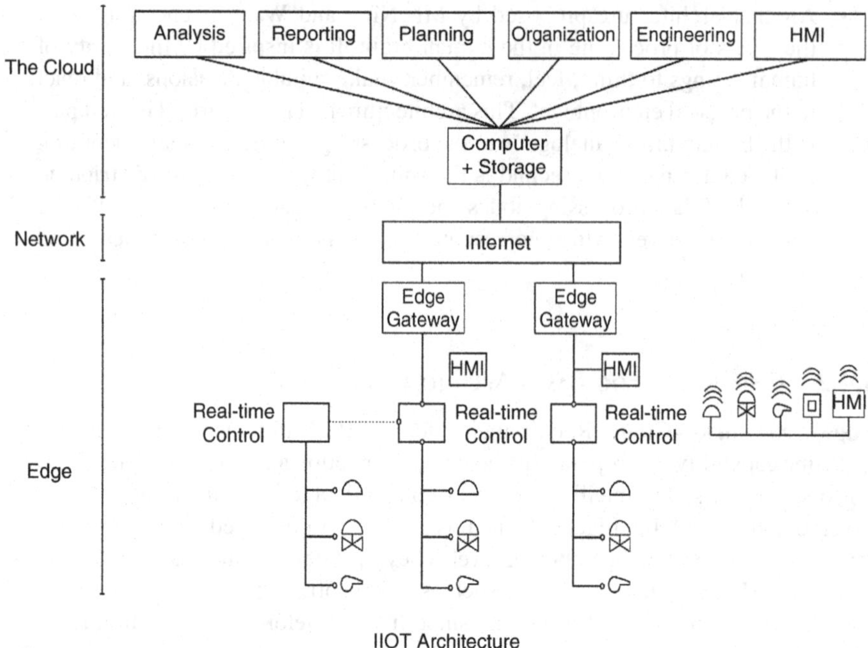

FIGURE 5.1 Flow layout of IIoT architecture.

to control outcomes in the real world through physical processes. What sets CPS apart from regular information and communications systems (IT or ICT) is the real-time character of their communications with the physical world. Both CPS and ICT systems process data and information; the focus of CPS is on controlling the physical process. CPS uses sensors to receive information about including measurement of physical parameters and actuators to engage in control over biological processes. CPS often involves a significant degree of autonomy. For example, CPS often can find whether to change the actuator's state or draw a human operator's awareness to some features of the environment being sensed.

5.4.2.2 Cloud Computing

IT services and assets can be uploaded to and retrieved from the Internet instead of directly connecting to cloud computing servers. It can store these files on cloud-based storage systems rather than local storage devices.

5.4.2.3 Edge Computing

A distributed computing paradigm always brings computer data storage closer to the location where it is required. In disparity to cloud computing, edge computing refers to decentralized data processing at the network's edge. The IIoT requires more of an edge-plus-cloud architecture than one based on the purely centralized cloud, to change productivity, products, and services in the industrial world.

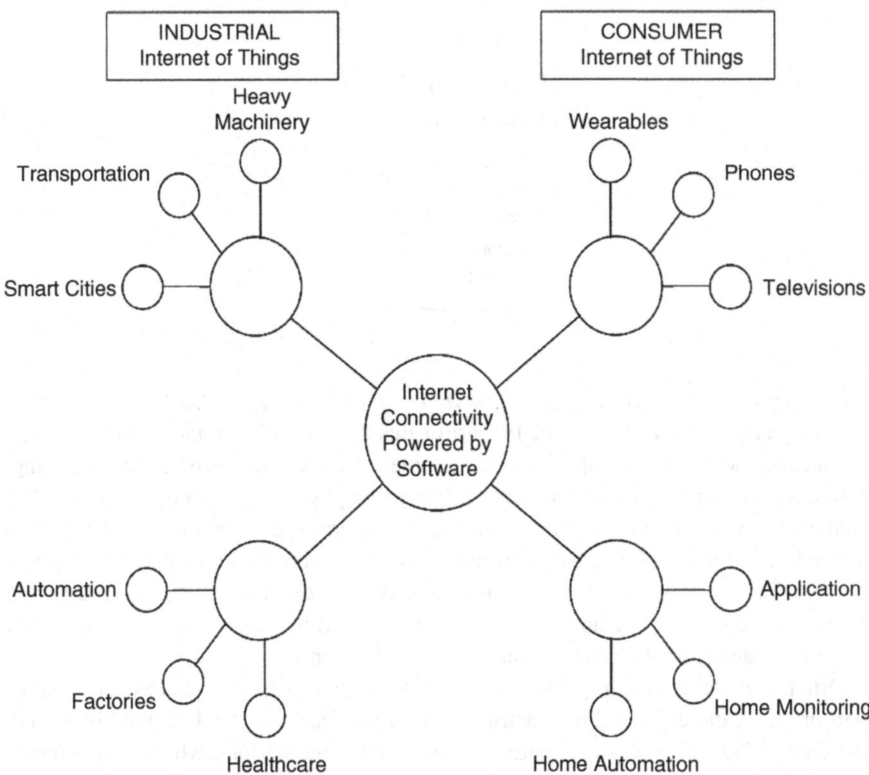

FIGURE 5.2 Flow logic diagram of IIoT vs IoT.

5.4.2.4 Big Data Analytics

It is the process of examining vast and varied datasets or big data.

5.4.2.5 AI and Machine Learning

AI is part of computer science in which intelligent machines have developed that work and behave like humans. ML is a core part of AI, allowing software to predict outcomes more accurately without programming.

There are mainly two kinds of systems architectures: cloud and fog computing. Note that this classification is quite different from the earlier classification, which was done based on protocols.

It is slightly vague about the nature of data generated by IoT devices and data processing. In some system architectures, cloud computers do the data processing in a significant centralized fashion. Such a cloud-centric architecture keeps the cloud at the center, applications above it, and the network of intelligent things below it. Cloud computing is censorious because it gives excellent flexibility and scalability, and it includes core infrastructure, platform, software, and storage services. Developers are offered storage tools, software tools, data mining, machine learning, and visualization tools through cloud computing.

TABLE 5.2
Layered Approach

Layered Approach
Transport Layer
Security Layer
Storage Layer
Reprocessing Layer
Monitoring Layer
Physical Layer

Another new architecture system, *known as fog computing*, is where the sensors and network gateways carry out a specific part of this data processing and analytics. A fog architecture shows a layered approach, as explained in Table 5.2, which has preprocessing, monitoring, storage, and security layers between the physical and transport layers. The monitoring layer also monitors power, resources, responses, and services. The temporary storage layer provides storage with the qualities having practical use, such as data duplication, distribution, and storage. Finally, the security layer executes encryption/decryption and protects data integrity and privacy. Monitoring and preprocessing are done on the network's edge before transferring data to the cloud system.

Often, the known "fog computing" and "edge computing" are used. The latter term predates the ester and is construed to be more generic. As described by *Cisco, Fog computing* refers to smart gateways and intelligent sensors, whereas *edge computing* is slightly more penetrative in nature. This paradigm involves implementing advanced data re-processing capabilities on physical devices such as motors, pumps, or lights. Usefulness provides more preprocessing of the available data in these devices, termed at the network's edge. The architectural diagram in Figure 5.1 remains largely unchanged in the context of system architecture. Consequently, it does not specifically depict edge computing.

Finally, the contrast between protocol and system architectures is not very clear. Often, the protocols and the system are code-signed. The generic five-layer IoT protocol stack (architectural diagram presented in Figure 5.1) will be used for fog and cloud architectures.

5.5 SOCIAL IOT

A new paradigm, social IoT (SIoT), considers social relationships between objects the same way humans form social connections. The three main facets of an IoT system are as follows:

- The SIoT is navigable. It can start with one device and navigate through all the devices that are connected to it. It is easy to locate new devices and services using such a social network of IoT devices.
- A need for trustworthiness (strength of the relationship) is present between devices (similar to friends on Facebook).

- It can use models identical to studying human social networks and examine IoT devices' social networks.

5.5.1 Basic Components

A typical social IoT setting can treat the devices and services as bots to set up relationships and modify them over time. It will allow the machines to cooperate and achieve a complex task seamlessly.

To make such a model work, many interoperating components are needed. Let us look at some of the significant elements in such a system.

5.5.1.1 ID

A unique method of object identification is needed. An ID can be assigned to an object based on traditional parameters such as MAC ID, IPv6 ID, a universal product code, or other custom methods.

5.5.1.2 Metainformation

Along with an ID, some metainformation about the device that describes its form and operation is needed. This is required to establish appropriate relationships with the device and appropriately place it in the universe of IoT devices.

5.5.1.3 Security Controls

It is similar to the "friend list" settings on Facebook. An owner of a device might limit the kinds of devices that can connect to it. These are typically referred to as *owner's controls.*

5.5.1.4 Service Discovery

This system is like a service cloud, where we need to have dedicated directories that store details of devices providing certain types of services. It becomes imperative to keep these directories up to date such that machines can learn about other devices.

5.5.1.5 Relationship Management

This module manages relationships with other devices. It also stores the instruments that a given device should connect with based on its services. For example, it makes a light brightness controller aware of a relationship with a light sensor.

5.5.1.6 Service Composition

This module carries the social IoT model to a new level. The final goal of designing such a system is to provide better-integrated services to consumers. For example, if a person had a power sensor with an air conditioner and this device sets up a relationship with an analytics engine, then the ensemble can yield much information about the usage patterns of the air conditioner. If the social model is very costly and there are many more devices, then it is feasible to compare the data with the usage patterns of other users and come up with even more high-quality information. For example, they can tell consumers that they are the largest energy consumers in their group or among their Facebook friends.

5.5.2 Representative Architecture of IoT

Most architectures proposed for the IoT have a server-side architecture as well. The network connects to all the interconnected components, aggregates the services, and acts as a single service point for users.

The server-side architecture typically has three layers. The first is the *base* layer containing a database that stores all the devices' details, attributes, metainformation, and relationships. The second layer (*component* layer) consists of code to interact with the devices, query their status, and use a subset to effect service. The uppermost layer is the *application* layer, which assists the users.

On the device (object) side, it broadly has two layers. The first is the *object* layer, which permits an instrument to join other devices, talk to them (via standardized protocols), and exchange details. The *object* layer passes details to the *social* layer. The social layer controls the execution of users' queries and applications and interacts with the application layer on the server.

IoT-enabled pumping systems enable industries to install a connected, flexible, and efficient pumping system. IIoT can transform the way industries work. It can create autonomous self-healing machines and enhance inventories using machine learning. Both philosophies have the same main character: intelligence, availability, and connected devices. The only difference between those two is their general utilization. While IoT is most commonly utilized for consumer usage, IIoT is used for industrial purposes such as manufacturing, supply chain monitoring, and management system. IIoT refers to IoT technology in industrial production. In the German-speaking world, IIoT is frequently referred to as Industry 4.0. Collated to the IoT in the private sphere, the focus, mainly in the industrial sector, is on the networking of machines (machine-to-machine) and seamless process chains. The basic concept of the IIoT is to integrate machine learning and big data technologies and thus considerably enhance the effectiveness of companies. However, complexity and requirements are much more prominent in IIoT than in IoT. IoT solutions comprise of multiple elements such as physical devices like sensors, actuators, and interactive devices; the network connecting these devices; the data collected from these devices and analyzed to create a meaningful experience; and last but not least, the physical context in which user interacts with the IoT.

5.5.3 The Four Elements of the Industrial Internet

The Industrial Internet of Things (IIoT) is a subset of the Internet of Things (IoT) used in industrial settings. IIoT systems have four main parts:

1. Intelligent assets: These are machines or other assets with sensors, processors, memory, and communications capabilities.
2. Data communications infrastructure: This infrastructure allows data to be communicated.
3. Analytics and applications: These interpret and act on the data.
4. People: People are also part of the IIoT system.

The sensing can be biometric, biological, environmental, visual, or audible (or all the above). IIoT machines must be capable of connecting and communicating in real time reliably. For industrial original equipment manufacturers (OEMs), this means including certified connectivity technologies that are more robust and using IoT embedded to handle and protect sensitive data at rest and in motion. IIoT is a network of intelligent devices connected to form systems that monitor, collect, exchange, and analyze data.

5.6 STANDARDS AND FRAMEWORKS OF IOT

IoT frameworks always help support the interaction between "things" and permit more complex structures like distributed computing and the development of distributed applications:

- IBM had earlier used cognitive IoT, which combines traditional IoT with machine intelligence and learning, contextual information, industry-specific models, and natural language processing.
- The XMPP Standards Foundation (XSF) is creating a framework called Chatty Things, a fully open, vendor-independent standard using XMPP to provide a distributed, scalable, and secure infrastructure.
- REST is a scalable architecture that allows things to communicate over Hypertext Transfer Protocol and is quickly adopted for IoT applications to speak from an item to a central web server.
- Message Queuing Telemetry Transport (MQTT) is a publish-subscribe architecture on top of TCP/IP, permitting bi-directional communication between a thing and an MQTT broker.
- Node-RED can be considered open-source software designed by IBM to connect application programming interfaces (APIs), hardware, and online services.
- Open Platform Communications (OPC) is a series of standards developed by the OPC Foundation system to connect computer systems to automated devices.
- The Industrial Internet Consortium, an Industrial Internet Reference Architecture, and the German Industry-4.0 are independent efforts to create a defined standard for IIoT-enabled facilities.

5.7 APPLICATIONS IN THE INDUSTRY

The term IIoT is often encountered in the manufacturing industries, referring to the industrial subset of the IoT. Significant benefits of the IIoT include improved productivity, analytics, and improvement in workplace transformation.

Establishing connectivity and data acquisition is imperative for IIoT; they are not the result but rather the foundation and path to something bigger. Predictive maintenance is an "easier" application of all the technologies, as it applies to existing assets and management systems. According to some studies, intelligent maintenance systems can reduce unexpected breakdowns and increase productivity, which is forecasted to save up to 12% time over scheduled repairs time, reduce overall maintenance time up to 30%, and reduce breakdowns up to 70%. CPS are the core technology of big industrial data, and they will be an interface between humans and the cyber world.

Integration of in situ sensing and actuation systems connected to the Internet can optimize energy consumption as a whole. It is expected to integrate IoT devices into all energy-consuming devices (switches, power outlets, bulbs, televisions, etc.) and effectively communicate with the utility supply company to balance power generation and energy usage. Besides home-based energy management, the IIoT is especially relevant to the Smart Grid. It furnishes systems to gather and act on energy and power-related data in an automated fashion to enhance the efficiency, reliability, economics, and sustainability of the production and distribution of electricity. Using advanced metering infrastructure gadgets connected to the Internet backbone, electric utilities can collect data from end-user connections and manage other distribution automation devices like transformers and reclosers.

5.8 IIOT FOR THE OIL AND GAS INDUSTRY

With the help of IIoT, large amounts of raw data can be stored and sent by the drilling rig operation and reservoir simulation modeling for cloud storage and analysis. With the implementation of IIoT technologies, the oil and gas industry can connect operating machines, devices, sensors, transmitters, and people through interconnectivity, which can help companies better address variation in demand, address cybersecurity, and minimize environmental impact. Across the supply chain, IIoT can also improve the maintenance process, overall safety, and connectivity. Drones can be used to detect possible oil and gas leaks in the processing system at an initial stage and at locations that are difficult to reach (e.g., remote areas). It can also identify weak spots in complex networks of oil and gas pipelines with built-in thermal imaging detection systems. Increased connectivity (data integration and communication) can help oil companies adjust oil production rates based on real-time data of inventory, storage, distribution pace, and forecasted demand. For example, one major oil company explained that by implementing an IIoT solution integrating data from multiple internal and external sources (such as work management system, process control center, pipeline control center attributes, risk scores, inline inspection findings, planned assessments, and leak history), it can monitor thousands of miles long pipelines in real time. IIoT allows monitoring pipeline threats perception in real time, improving risk management, and providing situational awareness. This type of benefit can also be applied to specific processes of the oil and gas industry. Oil and gas may be explored with 4D models built by seismic imaging/reservoir modeling. These models map the fluctuations in oil reserves and gas levels; they strive to point out the exact quantity of resources that have been exploited and forecast the lifespan of wells. The application of intelligent bottom hole sensors and automated drillers allows oil companies to drill a deepwater well to monitor geohazard well more efficiently.

Further, the product storing process can also be improved with the help of IIoT by collecting and analyzing real-time data to monitor the inventory levels and temperature control. IIoT can enhance the transportation process of oil and gas by implementing smart online sensors, meters, and thermal detectors to generate real-time geolocation data and monitor the products for safety reasons. These smart sensors can monitor the oil refinery processes and increase safety. Thus, with IIoT, the demand for products can be forecasted more accurately and automatically be communicated to the refineries and production installation to adjust production levels.

BIBLIOGRAPHY

Chernyshev, Maxim, Zubair Baig, Oladayo Bello, and Sherali Zeadally. Internet of Things (IoT): Research, Simulators, and Testbeds, *IEEE Internet of Things Journal*. 2017. 5 (3): 1637–1647.

Choudhury, Gopal, Manju Khari, and Mohamed Elhoseny. *Digital Twin Technology*, CRC Press, Boca Raton, FL. 2021. 48.

Sethi, Pallavi, and Smruti R. Sarangi. Internet of Things: Architectures, Protocol, and Application, *Journal of Electrical Engineering*. 2017: Article ID 9324035.

Umachandran, Krishnan, Igor Jurassic, Valentina Della Corte, and Debra Sharon Ferdinand-James. Industry 4.0: The New Industrial Revolution. In *Big Data Analytics for Smart and Connected Cities*, IGI Global, New York. 2019. 138–156.

Veneri, Giacomo, and Capasso Antonio. *Hands-On Industrial Internet of Things*, Packt, New York. 2018. 256.

BIBLIOGRAPHY

6 Artificial Intelligence and Robotics

6.1 INTRODUCTION

Artificial intelligence (AI) and robotics are part of digital technologies that will significantly impact the development of humanity soon. The ethics of AI and robotics is often focused on various critical sorts of new technologies. When it turns out to be rather quaint (trains are too fast for souls), some are predictably wrong when they suggest that technology will fundamentally change humans (telephones will destroy personal communication, writing will destroy memory, video cassettes will make going out redundant), and some are broadly correct but moderately relevant (digital technology will destroy industries that make the photographic film, cassette tapes, or vinyl records). Still, some are correct and deeply relevant (cars will kill children and fundamentally change the landscape). The purpose of describing the ethics of robotics in AI implementation is to analyze the issues and deflate the non-issues.

6.2 SINGULARITY AND SUPER-INTELLIGENCE

In some specific industries, the focus for current AI is thought to be an "artificial general intelligence" (AGI), contrasted to a technical or "narrow" AI. AGI is usually distinguished from traditional notions of AI as a general-purpose system and Searle's idea of "strong AI": computers given the right programs can be said to *understand* and have other cognitive states.

The idea of *singularity* is that if the trajectory of AI reaches up to systems with a human level of intelligence, then these systems would themselves have the ability to develop AI systems that surpass the human story of espionage, i.e., they are "super intelligent." Such super-intelligent AI systems would quickly self-improve or create even more intelligent systems. After reaching super-smart AI, this sharp turn of events is the "singularity" from which the development of AI is out of human control and hard to predict.

The fear that "the robots that will be created will take over the world" had captured human imagination even before computers.

Sometimes, ultra-intelligent machine can be defined as a machine that can far surpass all the intellectual activities of any human being, however clever he is, since the design of machines is one of these academic activities, an ultra-intelligent machine could design even better machines; there would then unquestionably be an "intelligence explosion," and the intelligence of human would be left far behind. Thus, the first ultra-intelligent machine is the last invention that man need ever make, provided that the machine is docile enough to tell us how to keep it under control.

DOI: 10.1201/9781003307723-6

6.3 STEPS FOR DESIGNING AN AI SYSTEM

The main steps for designing an AI system are as follows:

- Identify the actual problem
- Prepare the data list
- Choose the algorithms
- Train the algorithms
- Choose a particular programming language
- Run on a selected platform

There are three types of AI: narrow or weak AI, general or strong AI, and artificial super-intelligence. Currently, we have achieved only narrow AI. The same way AI solutions do. AI is one of the renowned technologies that are well known in the industry. There are four types of AI, which are as follows:

- Reactive machines
- Limited memory
- Theory of mind
- Self-awareness

The five major fields of robotics are human–robot interface, mobility, manipulation, programming, and sensors, and understanding these fields is important in robotics development. In order to have a more comprehensive understanding of robotics, robots can be categorized into five kinds:

- Cartesian
- Cylindrical
- SARA
- 6-Axis
- Delta

Each industrial robot type has specific elements that make them best suited for different applications. The main differentiators among them are their speed, size, and workspace, which are demonstrated by the multi-robot manufacturing network shown in Figure 6.1.

In the multi-robot manufacturing process, the interfacing program is parallelly connected to the interface circuit and controller, and it is activated by robotic sensors for generating/directing the main driver to manufacture/operate the robot arm motor and robot base, robotic gripper, and conveyor motor.

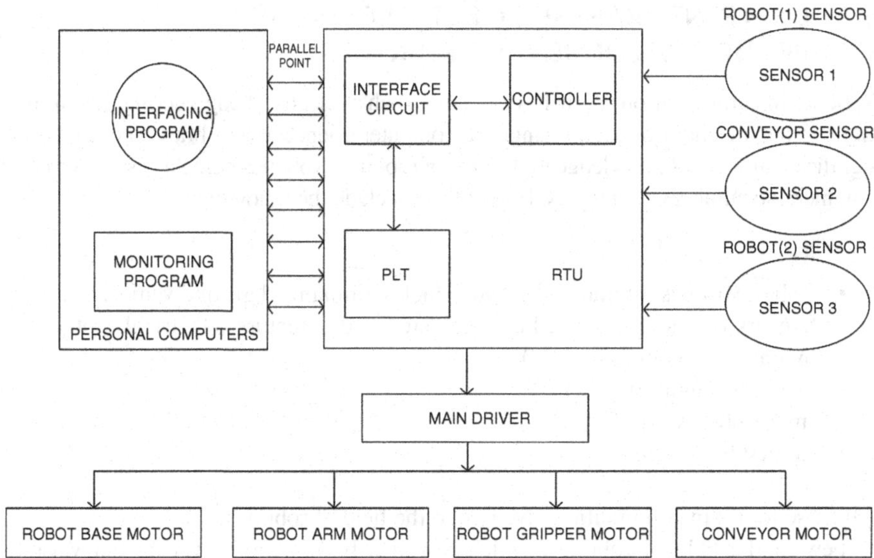

FIGURE 6.1 Flow logic of multi-robot manufacturing control network.

6.4 BASICS OF ROBOTICS

Robots spin wheels and pivot-jointed segments with some sort of actuator. Some robots use electric motors and solenoid valves as actuators, some use a hydraulic system, and some use a pneumatic system (driven by compressed gases). Robots use all these actuator types.

Currently, robotic engineers are designing the next generation of robots to look, feel, and act more humanlike, to make it easier for humans to warm up to a cold machine. Natural-looking hair and skin with embedded sensors will allow robots to react naturally in their environment.

Generally, there are five major types of robots given as follows:

- **Pre-programmed Robots:** Pre-programmed robots operate in a controlled environment where they perform simple, monotonous regular tasks.
- **Humanoid Robots**: Humanoid robots are robots that look like humans and mimic human behavior.
- **Autonomous Robots:** A journal focusing on the theory and applications of self-sufficient robotic systems.
- **Teleoperated Robots:** Robots that are controlled remotely by a human operator using a control station.
- **Augmenting Robots (also known as VR robots):** Robots that are designed to improve human capabilities or replace lost ones. The field of human augmentation combines biology and technology to improve people's lives.

6.5 DIFFERENT BRANCHES OF STUDIES INVOLVED IN THE DEVELOPMENT OF ROBOTICS

Robotics integrates various branches of studies such as electrical engineering, electronics engineering, mechanical engineering, and computer science. Every branch contributes a significant amount of knowledge to the area of robotics. Robotics being a vast area has an innumerable number of subtopics. Few of them include the following:

- AI
- Robotic drones (unmanned aerial vehicles, unmanned ground vehicles, etc.)
- Sensors (sound, pressure, humidity, etc.) and actuators (electrical motors, hydraulics, pneumatics, etc.)
- Speech recognition
- Image processing
- Embedded systems

All these topics in combination give rise to the field of robotics.

Few of the field of robotics, which can also be built for carrying out various operations, are as follows:

- Quadrotor
- Bridge laying robot
- Haptic robotic arm
- Multi-terrain robots
- Gesture controlled robot

The following are the leading engineering disciplines that are interlinked with each other in the designing of robots:

- **Mechanical Engineering:** Deals with the machinery and structure of the robots.
- **Electrical Engineering:** Deals with the controlling and intelligence (sensing) of robots.
- **Computer Engineering:** Deals with the movement development and observation of robots.

6.5.1 Major Role of Robots in Human Life

While AI can be entirely software, robots are physical machines that can move. Robots are subject to material impact, typically through "sensors," and they exert physical force onto the world, typically through "actuators," like a gripper or a turning wheel. Figure 6.1 shows the various manufacturing processes of robotic arm motors. Accordingly, autonomous cars or planes are robots, and only a minuscule portion of robots is "humanoid" (human-shaped), like in the movies. Some robots use AI, and some do not.

Typical industrial robots blindly follow completely defined scripts with minimal sensory input and no learning or reasoning (around 500,000 such new industrial robots are installed each year (IFR 2019 [OIR])). It is probably fair to say that while robotic systems cause more concerns in the general public, AI systems are more likely to have a more significant impact on humanity. Also, AI or robotic systems for a narrow set of tasks are less likely to cause new issues than more flexible and autonomous systems. Robotics and AI can thus be seen as covering three overlapping sets of plans: systems that are only AI, systems that are only robotics, and systems that are both.

R2-D2 robots can save human life because they get access to critical information (the number on that specific garbage disposal) at the right time. But robots and new technology already help save lives.

AI and robotics are powerful for automating tasks inside and outside the factory setting. In recent years, AI has become an increasingly common presence in robotic solutions, introducing flexibility and learning capabilities in previously rigid applications.

While AI is still in its nascent stage, it's been a transformative technology for some applications in the manufacturing sector, although many industries have yet to feel the impact.

6.5.2 Assembling a Robot

AI is a handy tool in robotic assembly applications. When combined with advanced vision systems, AI can help with real-time mid-course correction, instrumental in complex manufacturing sectors like aerospace. It can also help a robot learn on its own like which paths are best for specific processes while in operation. A simple robot sketch is shown in Figure 6.2.

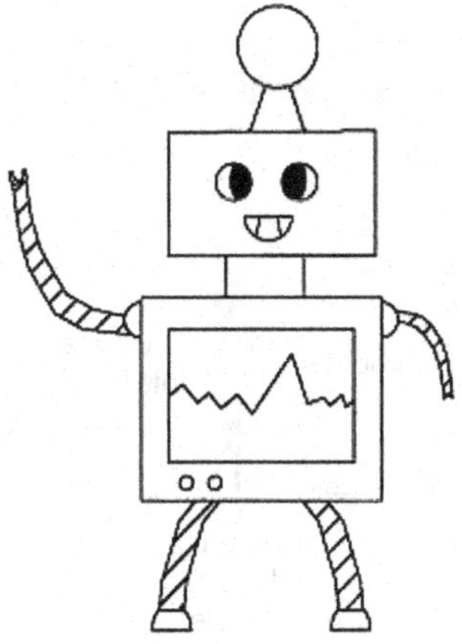

FIGURE 6.2 Sketch of a simple robot.

6.5.3 Packaging Industry Works by Robots

Robotic packaging uses forms of AI frequently for a quicker, less manual, and more accurate packaging. AI helps save certain motions a robotic system makes while constantly refining them, making installing and moving robotic systems easy enough for anybody to do.

6.6 OPEN SOURCE ROBOTICS

Very few robotic systems are now available as open-source systems with AI capability. This way, users can teach their robots to do custom tasks based on their specific application, such as small-scale agriculture. The convergence of open source robotics and AI could be a massive trend in the future of AI robots. When working together, robots are more intelligent, more accurate, and more profitable. The application of AI has developed at a much faster speed, but as it advances, so will robotics science.

6.7 ROBOTICS AND AI

The first thing to clarify is that robotics and AI are not the same. The two fields are entirely different.

A Venn diagram of the two fields is shown in Figure 6.3.

As can be seen from the figure, there is one small area where the two fields overlap: Artificially Intelligent Robots. It is within this overlap that people sometimes confuse the two concepts. These three terms relate to each other, and each of them is presented in the subsequent sections.

6.7.1 Robotics

Robotics is a branch of technology that deals with physical robots. Robots are programmable machines that can carry out a series of actions autonomously or semi-autonomously.

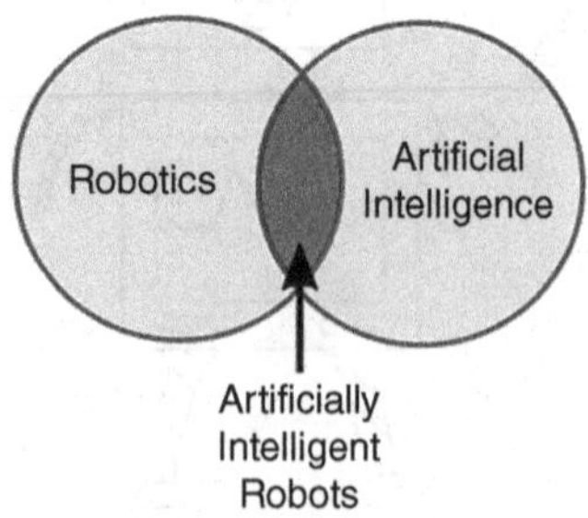

FIGURE 6.3 Venn diagram of two fields.

In my opinion, two critical factors constitute a robot:

- Robots interact with the physical world via sensors and actuators.
- Robots are programmable.

Robots are generally autonomous but some robots aren't. Tele-robots, for example, are entirely controlled by a human operator, but tele-robotics is still classed as a branch of robotics. This is one example where the definition of robotics is not very clear. It is surprisingly difficult to get experts to agree on precisely what constitutes a "robot." Some people say that a robot must be able to "think" and make decisions. However, there is no standard definition of "robot thinking." Requiring a robot to "think" suggests that it has some level of AI, but many non-intelligent robots that exist show that thinking cannot be a requirement for a robot.

However, when it is desired to define a robot, robotics involves designing, building, and programming physical robots that can interact with the physical world. Only a tiny part of robotics involves AI.

6.7.2 Basic Type of Cobot

A simple collaborative robot (cobot) is a perfect example of a non-intelligent robot.

For example, this can easily program a cobot to pick up an object and place it elsewhere. The cobot will then continue picking and placing objects precisely the same way until turning it off. This is an autonomous function because the robot does not require any human input after being programmed. The task does not require any intelligence because the cobot will never change what it is doing. Cobot is used in arranging auto storehouses.

6.7.2.1 A Last Caveat

The ethics of AI and robotics is a very new field within applied ethics, with significant dynamics but few well-established issues and no authoritative overviews—though there is a good outline (European Group on Ethics in Science and New Technologies 2018), and there are beginnings on societal impact.

Automation is driving a new way of working. The future workplace will feature a blend of the human and digital workforce ("bots"), and this symbiotic relationship will create many exciting possibilities to design. If not, now would be an excellent time to consider the jobs that will be emerging and to get trained and certified in the most in-demand and high-paying skills.

Robotics challenges AI by forcing it to deal with natural objects in the real world. Techniques and representations developed for purely cognitive problems, often in toy domain, do not necessarily extend to meet the challenge. Robots combine mechanical effectors, sensors, and computers. AI has made significant contributions to each component. AI helps save certain motions a robotic system makes while constantly refining them, making installing and moving robotic systems easy enough for anybody to do. Robots are now being used in a customer service capacity in retail stores and hotels around the world. AI is the simulation of human intelligence processes by machines, especially computer systems. These processes include learning, reasoning, and self-correction. Some of the applications of AI include expert systems, speech recognition, and machine vision.

6.7.3 Artificially Intelligent Collaborative Robots

This can extend the capabilities of a collaborative robot by using AI. When a camera is added to robot, robot vision comes under the category of "perception" and usually requires AI algorithms.

When desired, the cobot detects the object, picks up, and places it in a different location depending on the type of object. This would involve training a specialized vision program to recognize the different kinds of things. One way to do this is by using an AI algorithm called template matching.

In general, most artificially intelligent robots only use AI in one particular aspect of their operation. In this example, AI is only used in object detection. AI does not control the robot's movements (though the output of the object detector does influence its movements).

6.7.4 Industrial Robots

AI needs to be elaborated in conjunction with a robot. It involves developing computer programs to complete tasks that would otherwise require human intelligence. AI algorithms can tackle learning, perception, problem-solving, language understanding, and logical reasoning.

AI is used in many ways within the modern world. For example, AI algorithms are used in Google searches, Amazon's recommendation engine, and GPS route finders. Most AI programs are not used to control robots.

Even when AI is used to control robots, algorithms are only part of the more extensive robotic system, including sensors, actuators, and non-AI programming.

Often, but not always, AI involves some level of machine learning (ML), where an algorithm is "trained" to respond to a particular input in a certain way by using known inputs and outputs. Here, ML (discussed in Chapter 7) explains the difference between Robot Vision vs Computer Vision. The critical aspect that differentiates AI from more conventional programming is the word "intelligence." Non-AI programs simply carry out a defined sequence of instructions. AI programs mimic some level of human intelligence.

6.8 PURE AI FOR ALPHA GO

We can find one of the most common examples of pure AI in games. The classic example of this is the chess game, where the AI Deep Blue beat the world champion, Mr. Gary Kasparov, in 1997.

A more recent example is Alpha Go, an AI that beat Mr. Lee Sedol, the world champion Go player, in 2016. There were no robotic elements to Alpha Go. The playing pieces were moved by a human who watched the robot's moves on a computer screen.

6.9 AI-POWERED ROBOTS

AI-powered robots are equipped with sensors that provide data to analyze and act on in real-time. These sensors include cameras, vibration, proximity, accelerometers, and other environmental sensors.

AI algorithms are necessary when it is desired to allow the robot to perform more complex tasks.

A warehousing robot might use a path-finding algorithm to navigate warehouse management, like placement and sorting equipment at the proper location, cargo, and tools. A drone might use autonomous navigation to return home when it is about to run out of battery. A self-driving car might use a combination of AI algorithms to detect and avoid potential hazards on the road. These are all examples of artificially intelligent robots.

6.9.1 Application in Robotics and AI

Building on the advances made in mechatronics, electrical engineering, and computing, robotics is developing increasingly sophisticated sensorimotor functions that give machines the ability to adapt to their ever-changing environment. Until now, the industrial production system was organized around the machine; it is calibrated according to its environment and tolerating minimal variations. Today, it can be integrated more easily into an existing domain. A robot's autonomy in a domain can be subdivided into perceiving, planning, and execution (manipulating, navigating, collaborating). The main idea of converging AI and robotics is to optimize its level of autonomy through learning. This level of intelligence is measured as the capacity to predict the future, either in planning a task or interacting with the world. Robots with intelligence have been attempted many times. Although creating a system exhibiting human-like intelligence remains elusive, robots that can perform specialized autonomous tasks, such as driving a vehicle, flying in natural and artificial environments, swimming, carrying boxes and material in different terrains, picking up objects, and putting them down, do exist today.

Another important application of AI in robotics is the task of perception. Robots can sense the environment using integrated sensors or computer vision. In the last decade, computer systems have improved the quality of both sensing and image. Perception is essential for planning and creating an artificial sense of self-awareness in the robot. This permits supporting interactions of the robot with other entities in the same environment. This discipline is known as social robotics. It covers two broad domains: human–robot interactions (HCI) and cognitive robotics. The vision of HCI is to improve the robotic perception of humans, such as in understanding activities, emotions, non-verbal communications, and in being able to navigate an environment along with humans. The field of cognitive robotics focuses on providing robots with the autonomous capacity of learning and acquiring knowledge from sophisticated levels of perception based on imitation and experience. It aims at mimicking the human cognitive system, which regulates the process of developing knowledge and understanding through experience and sensitization; in cognitive robotics, some models incorporate motivation and curiosity to improve the quality and speed of knowledge acquisition through learning.

AI has continued beating all records and overcoming many challenges that were unthinkable less than a decade ago. Combining these advances will continue to reshape people's understanding of robotic intelligence in many new domains.

6.9.2 Software Robots

The term "software robot" refers to a type of computer program that autonomously operates to complete a virtual task.

One of the software robots examples is the search engine "bots" (aka "web crawlers"). These roam the internet, scanning websites and categorizing them for search.

6.9.3 Robotic Process Automation (RPA)

RPA, also known as software robotics, uses automation technologies to mimic back-office task of human workers, such as extracting data, filling in forms, and moving files. It combines API and user interface (UI) interactions to integrate and perform repetitive tasks between enterprise and productive application. RPA is a game-changing technology designed to automate high-volume, repeatable tasks that take up a large percentage of a workers' time. Robots are much like industrial robots that have automated blue-collar work in factories. RPA software mainly automates office work. This technology has misrepresented the word "robot" in the past few years.

RPA is a software technology that anyone can use to automate digital tasks. With RPA, software users create software robots, or "bots," that can learn, mimic, and execute rules-based business processes. RPA enables users to develop bots by observing human digital actions, show the bots what to do, and then let them do the work. RPA software bots can interact with any application or system the same way people do—except that RPA bots can operate around the clock, nonstop, much faster, and with 100% reliability and precision.

RPA bots have the same digital skillset as people—and some more. Think of RPA bots as a digital workforce that can interact with any system or application. For example, bots can copy-paste, scrape web data, make calculations, open and move files, parse emails, log into programs, connect to APIs, and extract unstructured data. And because bots can adapt to any interface or workflow, there's no need to change business systems, applications, or existing processes to automate.

RPA bots are easy to set up, use, and share. It will then be possible to set up RPA bots so that you can record video on your phone. RPA bots. It's as intuitive as hitting record, play, and stop buttons and using drag-and-drop to move files around at work. RPA bots can be scheduled, cloned, customized, and shared to execute business processes throughout the organization.

Robotic automation software provides a practical means of deploying new services in this situation, where the robots simply mimic the behavior of humans to perform the back-end transcription or processing. The relative affordability of this approach arises from the fact that no IT, new transformation, or investment is required; instead, the software robots simply leverage greater use of existing IT assets.

The hosting of RPA services also aligns with the metaphor of a software robot, with each robotic instance having its virtual workstation, much like a human worker. The robot uses keyboard and mouse controls to take actions and execute automation. Usually, these actions occur in a virtual environment and not on screen; the robot does not need a physical screen to operate; instead, it interprets the screen display electronically. The scalability of modern solutions based on architectures such as these owes much to the advent of virtualization technology. The scalability of large deployments would be limited by the available capacity to manage physical hardware and associated costs. The implementation of RPA in business enterprises has shown dramatic cost savings when compared to traditional non-RPA solutions.

There are, however, several risks with RPA. Criticism includes risks of stifling innovation and creating a more complex maintenance environment of existing software that now needs to consider using graphical user interfaces in a way they weren't intended to be used.

6.9.4 Chatbot

These are the programs that pop up on websites and talk to people with a set of pre-written responses.

Software bots are not physical robots. They only exist within a computer. Therefore, they are not real robots. Some advanced software robots may even include AI algorithms. However, software robots are not part of robotics engineering.

RPA is the technology that allows anyone today to configure computer software or a "robot" to emulate and integrate the actions of a human interacting within digital systems to execute a business process.

6.9.5 Role of RPA in Operations

According to most operations, groups adopting RPA have promised their employees that automation would not result in layoffs. Instead, workers have been redeployed to do more exciting work. One academic study highlighted that knowledge workers did not feel threatened by automation:

- The study embraced it and viewed the robots as teammates. The same research highlighted that, rather than resulting in a lower "headcount," deploy the technology in such a way as to achieve more work and greater productivity, and quicker output with the same number of people.
- Conversely, some analysts proffer that RPA represents a threat to the business process outsourcing industry. The thesis behind this notion is that RPA will enable enterprises to "repatriate" processes from offshore oil and gas locations into local data centers, benefiting from this new technology. The effect, if true, will be to create high-value jobs for skilled process designers in onshore

locations (and within the associated supply chain of IT hardware, data center management, etc.) but to decrease the available opportunity to low-skilled workers offshore.

6.9.6 RPA Actual Deployment

6.9.6.1 Enhancement of Greater Productivity by RPA Deployment

RPA bots create a step-change in employee productivity by accelerating workflows and enabling more work to get done by executing processes independently. In document-intensive industries like financial services, insurance, and the public sector, RPA bots can handle form filling and claims processing all hands-free.

RPA is deployed in the following primary sectors:

- Banking and Finance Process Automation
- Mortgage and Lending Process
- Customer Care Automation
- e-Commerce Merchandising Operation
- OCR Application
- Data Extraction Process

6.9.7 Harnessing AI Combined with RPA

When AI is combined with RPA to create Intelligent Automation, automating extends by order of magnitude and draws on the 80% of enterprise data that are unstructured. In procure-to-pay, invoice processing of non-standard vendor invoices is automated. In the insurance sector, extracting claims data and detecting potential fraud are automated. In human resource (HR), by understanding the employee's intent, the request intake is automated.

6.9.8 Effect on Society

Various studies carried out by multiple agencies, among other technological trends, are expected to drive a new wave of productivity and efficiency gains in the global labor market. Although not directly attributable to RPA alone, Oxford University conjectures that 35% of all jobs must be automated by 2035.

There are geographic implications to the trends in robotic automation. In the example above, where an offshore process is "repatriated" under the control of the client organization (or even displaced by a business process outsourcer from an offshore location to a data center), the impact will be a deficit in economic activity to the offshore site and a financial benefit to the originating economy. On this basis, in the developed economies, with skills and technological infrastructure to develop and support a robotic automation capability, the trend can be expected to benefit.

6.10 A TYPICAL MECHANICAL DESIGN OF A ROBOTIC ARM—A CASE STUDY

The mechanical design aspect of the robotic arm has to be considered for both design specification and design parameters. A robotic arm has certain design aspect specifications, and specific parameters are to be taken into consideration. Since the invention is an area related to thought, many plans come to mind at the initial stages of the program. Everything might be fruitful for designing robotic arm of 18 axis, and the trial-and-error method cannot be trusted blindly. So, keeping all these things in mind, a study has decided to design the robotic arm whose dimensions are loosely based on the dimension standards of the FANUC robotic arm. The essential points to be noted and followed for the design are as follows.

6.10.1 Functionality of a Robotic Arm

The arm should have the ability to lift, move, lower, and release an object while closely mimicking the motion of the human arm with full extension. Any device that can perform the required actions to pick and place an object required would have met the requirements of this criterion. The number of parts in this particular robotic arm is taken by comparing it with a human arm. Let the action of a human hand picking up a container appear in mind. Humans have the waist, shoulder, elbow, arm, wrist, and fingers to do the job. This is the motivation for the choice of the number of parts. This robotic arm also has five parts and five joints that are pretty much like the human hand.

6.10.2 Reliability of a Robotic Arm Design

The device should be able to pick up and smoothly place objects consistently, i.e., the device's motion should be smooth enough not to drop the things lifted. Therefore, any device that can lift and move an object from one place to another without losing any grip would meet the criteria. After a detailed study, the choice of the end effector is made. Since this device is used to pick and place metal sheets, the first common thought any mind would get is that a magnet can lift the sheet. But the problem with that is that the sheet's thickness is so small that there is a very high chance of more than one sheet being picked. Suppose more than one sheet is fed to the shearing machine at a time, it hurts the shearing blade badly, reducing the blade's life. The next option in front of this work was to use suction cups to lift the sheets. It is the most commonly used technique for transporting metal sheets in industries all over the world. So, it is necessary to use this technique for this purpose. The industry also used suction cups and a linear robot (conveyor) to transport sheets in the printing process.

6.10.3 Robotic Motion Range and Speed

Like the human body, robots are constructed with the same joints between bones; here, they had a constrained limit for the movement of the axis. In this design application, every particular axis has its capacity for motion. The degree of movement of the robot is calibrated from the center base of 19 axes.

By this, the speed in pick and place operation might vary, and this occurs because each axis moves at different rates. The sweeping motion of the process is recorded in terms of degrees traveled per second.

6.10.4 Payload Consideration for a Robot

The payload of each robot is determined by its slight weight, so critical specifications and tooling weights are sorted out. This information is useful for categorizing robots based on these specifications.

6.10.5 Reach of the Robot

This articulated robot needs to check the two extremities that are nothing but the V-reach and H-reach. Vertical reach is considered to know how high our robot can go in terms of height extension. At the same time, the horizontal space is evaluated to see the distance of a fully extended arm from base to wrist. In a few other applications, even a horizontal short reach needs to be considered.

6.10.6 Axes of the Robot

The distinctive segments of this robot are associated with mechanical joints that serve as an axis of movement. It has been designed for an articulated robot with a five-axis of action. Generally, the knowledge of industrial robots is designed to have six-axis of movement, but the number and placement of robots just give flexibility variation for each model.

6.10.7 Design of a Robot—Part 1

This part is designed first because, based on its measurements, the measurement of remaining parts could be done. And the weight of all other elements, including the payload, will significantly affect this part since it is the base of this robotic arm. This part is an assembly of two different parts. Part 1 must rotate on its axis, and other parts are connected to this. So, it is the primary source of transportation. The two subparts in this assembly are the upper part and a shaft. The upper part is designed such that its bottom is one side of an Oldham coupling, as shown in Figure 6.4.

The upper part's dimensions are tabulated in the later part of the text. This part stays in the upper area of the base and will be visible. A shaft and a key are attached to this part from the bottom; a power source is connected from here and is made to rotate. It is supposed to be the component that transfers motion (rotational) from the power source to the upper part of the body. The shaft and key and the whole assembly of Part 1 are shown in Figure 6.5. The shaft and key inserted in this assembly act like a typical Oldham coupling. A general Oldham coupling has three flanges, one coupled to the input, one connected to the output, and a middle disc joined to the first two by tongue and groove. The tongue and groove on one side are perpendicular to

FIGURE 6.4 Assembly of Part 1.

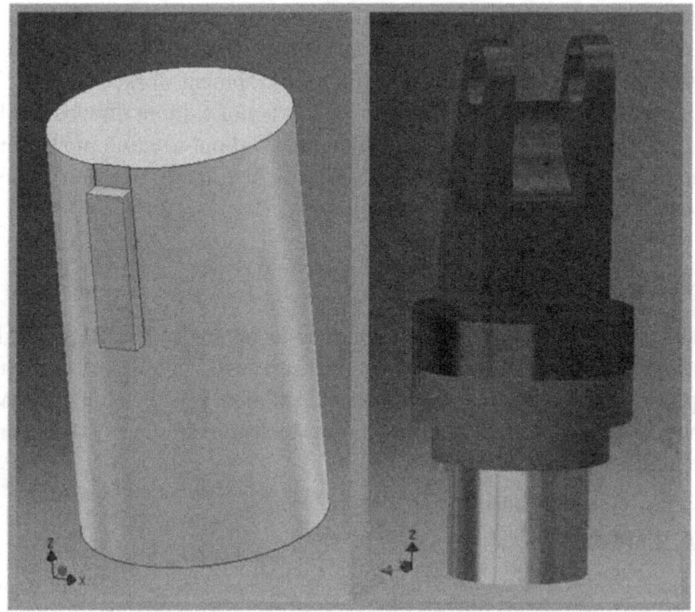

FIGURE 6.5 The shaft and key of Oldham coupling.

the tongue and groove on the other side. The middle flange rotates around its center at the same speed as the input and output shafts. Its center traces a circular orbit, twice per rotation, around the midpoint between input and output shafts.

For this operation, modification of the Oldham coupling a bit is done. Instead of using three flanges, only two are used, which are not flanges exactly. We designed the ends of the two parts in the base as Oldham couplings, as shown in Figure 6.5.

Since this is a pick and place operation, two discs instead of three might withstand the torque ranges of this operation. As can be seen from the Figure 6.5, the lower part of Oldham coupling has a shaft and a key. The upper end of the collars has a flange designed directly with the shoulder of the robotic arm. All the assembly described till now will be assigned a rotatory motion. That means a motor is attached to the shaft at the lowermost part as shown in Figure 6.5, and if it rotates, the whole robotic arm rotates.

The dimensions of the shaft are given as follows:

Length of the shaft (L) = 0.182 m
The radius of the shaft (r) = 0.12 m
Volume of the shaft (V) = 0.002055 m^3

6.11 ROBOTIC PROCESS AUTOMATION-2.0

RPA 2.0, often referred to as "unassisted RPA" or RPAAI, is the next generation of RPA-related technologies. Technological advancements and improvements around AI technologies are making it easier for businesses to take advantage of the benefits of RPA without dedicating a large budget for development work.

While unassisted RPA has several benefits, it is not without drawbacks. Utilizing unassisted RPA can run a process on a computer without needing input from a user, freeing up that user to do other work. However, to be effective, obvious rules need to be established for the processes to run smoothly.

6.12 HYPERAUTOMATION

Hyperautomation is the application of advanced technologies like RPA, AI, ML, and process mining to augment workers and automate processes in ways that are significantly more impactful than traditional automation capabilities. It is the combination of automation tools to deliver work. It requires various tools to help support replicating pieces where the human is involved in a task.

BIBLIOGRAPHY

Corke, Peter. *Robotics, Vision & Control*. Springer, Berlin, Heidelberg, 2011 September 5. 43–59.

David, Mathew. "RPA Fundamentals: Getting Started with Robotic Process Automation." Simplilearn. 2021 April 28. https://www.simplilearn.com/rpa-fundamentals-article.

Govers Francis X. *Artificial Intelligence for Robotics*. Packt, Birmingham, United Kingdom, August 2018. 43–58, 70–104.

Mahdy, Qaysar Salih, Ghani Hashim, Idris Haldi Saleh, and Ganesh Babu Loganathan. Evaluation of Robot Professor Technology in Teaching and Business, *IT in Industry*. 2021 March 18. 9: 1182–1194.

Robotics Online Marketing Team. "How AI Is Used in Today's Robots." Telefonica, Germany, RIA Blog. 2018 November 9. https://www.automate.org/blogs/how-artificial-intelligence-is-used-in-today-s-robots.

7 Machine Learning and Artificial Intelligence

7.1 INTRODUCTION

The application of artificial intelligence (AI) techniques in science and engineering was met with mixed reactions. It meant different things to different people. To the pessimists and antagonists, it is perhaps another fairy tale or fiction in which computers are presented to be able to "develop the ability to think and feel and take over the world in some dystopian future." It is simply treated with underserved disbelief, distrust, and total despair.

Machine learning (ML) is a scientific aspect of AI concerned with designing and developing suitable algorithms that allow computers to learn based on data, such as from logs or core datasets. A primary focus of ML research is to automatically learn to recognize complex patterns and make intelligent decisions based on data. Hence, ML is closely related to statistics, probability theory, data mining (DM), pattern recognition, AI, adaptive control, and theoretical computer. The rapid increase in the availability of computers and data has led to the increased prominence of ML.

7.2 DATA SCIENCE AND AI

When exposed to new data, ML applications learn, grow, and develop by themselves. In ML, computers generally locate the insight information without being communicating where to look for it. ML does this by leveraging algorithms that learn from data in an iterative process. ML is a part of exciting subsets of AI. It's essential to understand what makes ML work and, thus, how it can be used in the future. Three primary processes are involved in the knowledge acquisition after identifying the sample problem through the interaction with the experts in the problem domain:

- Formulating the concept
- Sample problem implementation
- Knowledge representation developments

The sources of knowledge, include experts, textbooks, reference manuals or monographs, databases, reports environment videotape, etc. The different types of expertise may be numerical data, simple propositions, facts and figures, rules, procedures, concepts, and formulas, i.e., mathematical relations.

ML involves computers discovering the way to perform various tasks without being explicitly programmed to do so. This involves computers learning from data provided so that they can carry out specific tasks. For simple tasks assigned to computers, it may be possible to program algorithms telling the machine way to execute all steps required

DOI: 10.1201/9781003307723-7

to solve the problem at hand; no learning is needed on the computer. Advanced tasks will be challenging for a human to create the necessary algorithms manually. In practice, it can be more effective to help the machine develop its algorithm, rather than having human programmers specify every needed step.

7.3 MACHINE LEARNING

Powered by many computer servers and ability to process billions of data points in real time, ML algorithms can layer information associated with multiple variables on top of one another. This allows for fast identification of trends and patterns that otherwise is time-consuming to detect, even to the most experienced human eyes.

Analytical software provides visual views of data to help and turn it into actionable information. See, for instance, instead of monitoring a single variable like pressure differential in a reservoir, ML software can consider several factors critical to the overall drilling strategy. These include equipment ratings, seismic vibrations, rock permeability, and geothermal gradients. When layered, these data can be used to determine the optimal direction of the drill bit and how it should be controlled (i.e., rate of penetration) as it drills through the rock layer.

Information is logged, contextualized, and visualized via human–machine interfaces, allowing personnel to monitor the overall performance of oil wells and make more intelligent decisions aimed at improving operations.

Predictive software can also analyze data to determine if downhole well conditions are conducive to potentially catastrophic events such as lost circulation, stuck drill pipe, or blowouts. By leveraging data to understand the degree contributes to the similar of such an event, algorithms can provide recommendations to the control system and operating personnel to reduce the odds of such an occurrence. This has become critical, given that the expenditure of non-productive time can be very high per day in some cases.

7.4 DECOMPOSITION OF MACHINE LEARNING

ML is not magical pixie dust. ML cannot simply automate all decisions through data. This had constrained by available data and the models that were used. ML models are relatively straightforward function mappings that include characteristics such as smoothness. With some notable deviations, e.g., speech and image data, inputs are constrained in the form of vectors, and the model consists of a mathematically well-behaved function, to be put into the right sub-process to automate ML.

Any job happening again is a candidate for automation, but many of the repetitive tasks performed as human's brain are more complex than any individual algorithm can replace. The selection of which scheme to automate becomes critical and has downstream effects on the overall system design.

7.5 MACHINE LEARNING PROCESS AND APPLICATION SOFTWARE

ML application processes are connected with the algorithm, which knows the rules from experience as below:

- Speech recognition
- Labeled examples
- User feedback
- Surrounding environment (self-driving car)

ML brings together statistics and computer science and also identifies the data. Consider the following groups:

- Statistical method
  - Infer conclusion from data
  - Estimate reliability of prediction
- Computer science
  - Large-scale computing architecture
  - Algorithm for capturing, manipulating, indexing, combing, retrieving, and performing prediction on data
  - Software for pipeline operation that manages the complexity of multiple set back in the operation
- Economics, biology, and psychology
  - How the performance of one individual or a system can improve their performance

  ML is also used for fraud detection of credit scores, spell checking, Web search, query, etc.
- Model learning algorithms

  ML tools are algorithmic applications of AI that generate the systems's ability to learn and improve without ample human input; similar concepts are DM and predictive modeling.

7.5.1 Machine Learning Operating Software

Python and AI: Python is one of the specific software programming languages used by various software developers. It was created in 1991 and, since its inception, has been one of the most widely used languages along with C++, Java, etc.

Being one of the best-known programming languages for AI and neural networks, Python has taken a big lead. AI, in combination with h Python, is one of the best tools in the operating software system.

Python, with its rich technology competency, has an extensive set of libraries for AI and ML. Some of them are Keras, TensorFlow, and Scikit-learn for ML, and NumPy for high-performance scientific computing and data analysis software. Python is a more superior language over C++ for AI. C++ is a lower-level language and requires more experience and skill to master. ML can provide capabilities that can be used at any step in the process of working through an ML problem.

A comparison chart (Table 7.1) of the four major ML software platforms is written in language to be corrected.

There are eight popular ML software, which are as follows:

- Tensor Flow: The name for ML in the Data Science industry is Tensor Flow.
- Shogun: Shogun is a popular, open-source machine learning software.

- Apache Mahout
- Apache Spark MLlib
- Oryx 2
- H20.ai
- Pytorch
- RapidMiner

TABLE 7.1
Machine Learning Comparison Chart

Platform		
Scikit-Learn	Linux, Mac OS, Windows	Python, Cython, C, C++
Py Torch	Linux, Mac OS, Windows	Python, C++, CUDA
Tensor Flow	Linux, Mac OS, Windows	Python, C++, CUDA
Weka	Linux, Mac OS, Windows	Java

7.5.2 Advantages of Python

Python is an interpreted language which, in common man's terms, means that it does not need to be converted into machine language instruction before execution and can be used by the developer directly to run the program.

Python is also a software programming language, which means it can be used across specific domains and technology.

Python also features a dynamic-type system and automatic memory management supporting system, and a wide variety of programming paradigms, including object-oriented, imperative, functional, and procedural, to name a few.

Python is available for all operating systems and also has an open-source offering titled CPython, which is garnering widespread popularity as well.

7.5.3 AI and Python

The reasons for choosing Python for AI over other languages are as follows:

Python offers the minor code among others and is, in fact, 1/5 the number compared to other OOP languages.

Python has prebuilt libraries like Numpy for scientific computation, SciPy for advanced computing, and PyBrain for ML (Python ML), making it one of the best languages for AI.

Python developers around the world provide comprehensive support and assistance via forums and tutorials, making the job of the coder easier than any other popular languages.

Python is the most flexible of all others, with options to choose between the OOPs approach and scripting. It can also use integrated development environment (IDE) itself to check for most codes and is a boon for developers struggling with different algorithms.

7.5.4 Decoding Python Alongside AI

Python, along with packages like NumPy, Scikit-learn, iPython Notebook, and matplotlib, forms the basis to start an AI project.

NumPy is used as a container for generic data comprising of an N-dimensional array object, tools for integrating C/C++ code, Fourier transform, random number capabilities, and other functions.

Another helpful library is Pandas, an open-source library that provides users with easy-to-use data structures and analytic tools for Python.

Matplotlib is another service that is a 2D plotting library creating publication-quality figures. This can use matplotlib to up to six graphical user interface toolkits, web application servers, and Python scripts.

The next step will be to explore k-means clustering and knowledge about decision trees, continuous numeric prediction, and logistic regression.

Some of the most commonly used Python AI libraries are AIMA, pyDatalog, SimpleAI, and EasyAi. There are also Python libraries for ML like PyBrain, MDP, scikit, and PyML.

7.5.5 Python Libraries for General AI

AIMA: Python implementation of algorithms from Russell and Norvig's *Artificial Intelligence: A Modern Approach.*

pyDatalog: Logic Programming engine in Python.

simple: Python implementation of many of the AI algorithms described in the book *Artificial Intelligence: A Modern Approach*. It focuses on providing an easy-to-use, well-documented, and tested library.

EasyAI: Simple Python engine for two-player games with AI (Negamax, transposition tables, game solving).

7.5.6 Python for Machine Language

Python and the various libraries it offers are used for ML.

PyBrain: It is a flexible, simple yet effective algorithm for ML tasks. It is also a modular ML library for Python providing a variety of predefined environments to test and compare algorithms.

Pym: It is a bilateral framework written in Python that focuses on support vector machines (SVMs) and other kernel methods. It is supported on Linux and Mac OS X.

Scikit-learn: It is an efficient tool for data analysis while using Python. It is an open-source and the most popular general-purpose ML library.

MDP-Toolkit: It is another Python data processing framework that can be easily expanded; it also has a collection of supervised and unsupervised learning algorithms and other data processing units that can be combined into data processing sequences and more complex feed-forward network architectures. The implementation of new algorithms is easy and intuitive.

The base of available algorithms is steadily increasing and includes signal processing methods (Principal Component Analysis, Independent Component Analysis, and Slow Feature Analysis), manifold learning methods ([Hessian] Locally Linear Embedding), several classifiers, probabilistic methods (Factor Analysis, results-based management (RBM)), data preprocessing techniques, and many others.

7.5.7 Python Libraries for Natural Language and Text Processing

Natural Language Toolkit (NLTK): Open-source Python modules, linguistic data, and documentation for research and development in natural language processing and text analytics. Distributions are available for Windows, Mac OSX, and Linux.

7.5.8 Comparison between Python and C++ for AI

Python is more popular than C++ for AI, with 57% of developers preferring it. This is because Python is easy to learn and use, and it has many libraries for data analysis. However, C++ outperforms Python in terms of performance because it is a statically typed language, eliminating typing errors during runtime and creating more compact and faster runtime code.

Python's dynamic nature reduces complexity when collaborating, allowing functionality implementation with less code. In contrast, C++ relies on powerful compilers that do specific optimization and can be platform-specific. Python code can run on any platform without needing specific configurations. Additionally, Python has an advantage in GPU-accelerated computing with Libraries such as CUDA Python and cuDNN offload primary computing for ML workloads to GPUs. This advantage makes any performance edge that C++ holds increasingly irrelevant.

Python also wins over C++ in simplicity, especially among new developers. C++ is a lower-level language and requires more experience and skill to master. Python's simple syntax allows for a more natural and intuitive ETL process and faster development, enabling developers to test ML algorithms without implementing them quickly.

In summary, Python's simplicity, readability, and thriving community, supported by collaborative tools like Jupyter Notebooks and Google Colab, give it the edge over C++ for AI.

7.5.9 Case Study

An experiment to bring AI to use with an Internet of Things was done to make an IoT application for employee behavioral analytics. The software provides valuable feedback to employees through employee emotions and behavior analysis, thus enhancing positive changes in management and work habits.

With Python ML libraries OpenCV and Haar Cascading concepts for application training, a sample proof of concept (POC) was built to detect basic emotions like happiness, anger, sadness, disgust, suspicion, contempt, sarcasm, and surprise through wireless cameras attached at various bay points.

The data collected were fed to a centralized cloud computing database where daily emotional quotient within the bay or even the entire office could be retrieved at the click of a button either through an android device or desktop.

Developers are trying gradual progress in analyzing further complex points on facial emotions and mine more details with the help of deep learning algorithms and ML, which can help analyze individual employee performance and support proper employee/team feedback.

7.6 MACHINE LEARNING CLASSIFICATION

ML approaches are traditionally divided into three broad categories, depending on the nature of the "signal" or "feedback" available to the learning system.

Other approaches have been developed, which don't fit neatly into this three-fold categorization, and sometimes, more than one category are used by the same ML system—e.g., topic modeling, dimensionality reduction, or meta-learning.

7.6.1 Machine Learning Categories

The three standard algorithms of ML are Supervised Learning, Unsupervised Learning, and Semi-supervised Learning.

7.6.1.1 Supervised Learning

Supervised learning is a type of ML technique in which the algorithm generates a function that maps input to the desired outputs with the slightest error. Examples are the prediction of permeability and the identification of lithofacies and rock types.

Two of the most common supervised ML tasks are classification and regression. In classification problems, the machine must learn to predict discrete values. That is, the machine must expect the most probable category, class, or label for new examples. Applications of classification include predicting whether a stock's price will rise or fall or deciding if a news article belongs to the politics of leisure section. In regression problems, the machine must predict the value of a continuous response variable. Examples of regression problems include predicting the sales for a new product or the salary for a job based on its description.

7.6.1.2 Unsupervised Learning

Unclassified learning is the ML technique in which a set of inputs are analyzed without the target output and unlabeled data, and the system attempts to uncover the patterns from the data. There is no label or target given for the examples. One common task is to group similar models together, called clustering. Without the aspect of known data, the input cannot be guided to the algorithm, which is where the unsupervised term originates from. Data are fed to the ML algorithm and are used to train the model. The trained model tries to search for a pattern and gives the desired response. In this case, it is often like the algorithm is trying to break code like the Enigma machine, without the human mind directly involved but rather a machine.

The top algorithms currently being used for unsupervised learning are as follows:

1. Partial least squares
2. Fuzzy means
3. Singular value decomposition
4. K-means clustering
5. Apriori
6. Hierarchical clustering
7. Principal component analysis

Unsupervised learning algorithms take a set of data that contain inputs and fine structure, like grouping or clustering of data points. The algorithms, therefore, learn from test data that have not been labeled, classified, or categorized. Instead of responding to feedback, unsupervised learning algorithms identify commonalities in the data and react based on the presence or absence of such commonalities in each new piece of data. A central application of unsupervised learning is in the field of density estimation in statistics, such as finding the probability density function. However, unsupervised learning encompasses other domains involving summarizing and explaining data features.

Cluster analysis is the assignment of a set of observations into subsets (called clusters) so that comments within the same cluster are similar according to one or more predesignated criteria, while statements drawn from different clusters are dissimilar. Other clustering techniques make different assumptions on the structure of the data, often defined by some similarity metric and evaluated, for example, by internal compactness, or the similarity between members of the same cluster, and separation, the difference between clusters. Other methods are based on estimated density and graph connectivity.

7.6.1.3 Semi-Supervised Learning

Semi-supervised learning can be the group in between unsupervised learning (without any labeled training data) and supervised learning (with completely labeled training data). Most of the training examples are missing training labels; still many machine learning researchers have found that unlabeled data, when used in conjunction with a small amount of labeled data, can produce a considerable improvement in learning accuracy.

In supervised learning, which is the weak in nature, the training labels may be noisy, limited, or imprecise; however, these labels are often cheaper to obtain, resulting in larger practical training sets.

7.6.2 Reinforcement Learning

A computer program tries to interact with a dynamic environment in which it must perform a particular target (such as driving a vehicle or playing a game against an opponent). As it navigates its problem space, the program is provided feedback that's analogous to rewards, which it tries to machine. Reinforcement learning is also referring to target-oriented algorithms, which learn how to attain a complex objective (goal) or maximize along a particular dimension over many steps. This method allows machines and software agents to automatically determine the ideal behavior within a specific context in order to optimize its performance. Simple reward feedback is required for the agent to learn which action is best; this is known as the reinforcement signal, for example, maximize the points won in a game over many moves. Like traditional types of data analysis, here, the algorithm discovers data through a process of trial and error and then decides what action results in higher rewards. Three major components make up reinforcement learning: the agent, the environment, and the steps. The agent (discussed in Chapter 2) is the learner or decision-maker, the environment includes everything that the agent interacts with, and the actions are what the agent does.

Reinforcement learning occurs when the agent chooses actions that maximize the expected reward over a given time. This is the easiest way to achieve when the agent is working within a sound policy framework.

The rapid evolution in ML has caused a subsequent rise in the use cases, demands, and the sheer importance of ML in modern life. Big Data has also become a well-used buzzword in the last few years. This is, in part, due to the increased sophistication of ML, which enables the analysis of large chunks of Big Data. ML has also changed the way data extraction and interpretation are made by automating generic methods/algorithms, thereby replacing the traditional statistical techniques.

To get the most value out of Big Data, other ML tools and processes that leverage various algorithms include the following:

- Comprehensive data quality and management
- GUIs for building models and process flows
- Interactive exploration data and visualization of geo-model results and prediction
- Comparisons of different ML models to quickly identify the best one object (sweet spot in oil reservoir) for drilling
- Automated ensemble model evaluation to determine the best performers
- Easy model deployment so that it can get repeatable, reliable results arrived quickly
- An integrated end-to-end platform for the automation of the data to decision process

Reinforcement learning is an area of ML concerned with how software agents ought to take actions in an environment so as to maximize some notion of cumulative reward. Due to its generality, the field is studied in many other disciplines, such as game theory, control theory, operations research, information theory, simulation-based optimization, multi-agent systems, swarm intelligence, statistics, and genetic algorithms. In ML, the environment is typically represented as a Markov decision process (MDP). Many reinforcement learning algorithms use dynamic programming techniques. Reinforcement learning algorithms do not assume knowledge of an exact mathematical model of the MDP and are used when accurate models are infeasible. Reinforcement learning algorithms are used in autonomous vehicles or in learning to play a game against a human opponent.

7.6.3 Deep Learning Algorithms

In AI, often heard concepts are ML and deep learning. In fact, these have an inclusive relationship, and deep learning is a specific form of learning in ML. It is mainly based on algorithms of neural networks. At present, deep learning has made significant progress in fields of image recognition, speech recognition, natural language processing, audio recognition, social network filtering, machine translation, medical image analysis, and board game programs.

7.6.4 Artificial Neural Network (ANN)

The development of ANN has been inspired by the workings of the human brain; as such, there have been many attempts to artificially simulate the biological process that leads to intelligent behavior. ANN is a close emulation of the biological nervous system. A simplified diagram of a neuron can be drawn. Just as a biological neuron acts as an integrator of the multiple excitatory and inhibitory inputs it receives at cell body and dendrites, combing all these sums of inputs and excitations; and forwarding the result down the axon to other neurons and muscles, in the form of a pulse, the ANN's neuron multiplies the inputs by weights to calculate the sum. It applies a threshold and transmits the result of the computation to the subsequent neurons. Basically, ANN has been generalized to mathematical function T2FLS (type-two fuzzy logic system, which was introduced by Mr. Zadeh), as in the following equation:

$$\mathrm{Yi} = \mathrm{f}\left(\Sigma\, \mathrm{Wik\ Xk} + \mu\mathrm{i}\right), \tag{7.1}$$

where Xk is the input to the neuron I; Wik are weights attached to inputs, μi is the threshold, offset, or bias; f (o) is a transfer function, and Yi is the output of the neuron. The transfer function f (o) can be any linear, non-linear, price-wise linear, sigmoidal, tangent hyperbolic, and polynomial function. T2FLS structures are shown in Figure 7.1, which is a type-2 grade of membership. It is similar to the T1FLS structure except that the outprocessing block contains the fuzzified system.

When it comes to deep learning algorithms, one needs to know about convolutional neural networks (CNNs) and recurrent neural networks. The neural network is similar

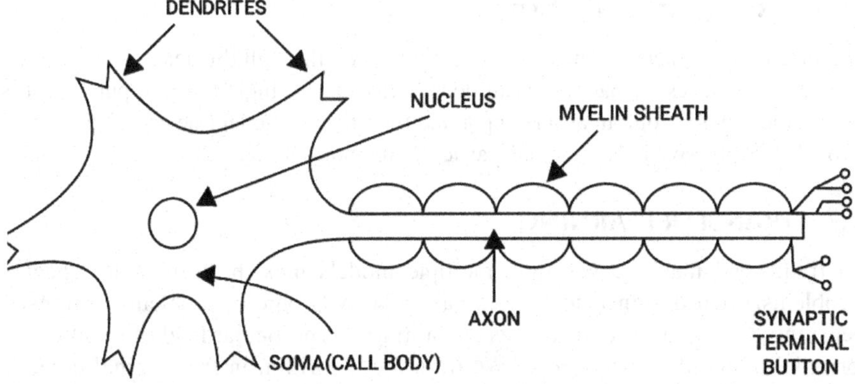

FIGURE 7.1 The structure of T2FLS—a biological neuron.

to the neural transmission of the human brain, from one input unit to the next input unit to get a result. This is the principle of a simple neural network, which is to simulate the transmission of information from nerves in the human brain. It transfers data from one neuron to another and then passes down. With the rise of neural network algorithms, deep learning algorithms can solve many practical problems.

7.6.5 BP Neural Network

After the invention of the neural network algorithm, many problems have been solved to a certain extent. At the same time, people are constantly optimizing this algorithm. First, a very widely used and classic one is the back propagation (BP) neural network. BP neural network has one or more hidden layers than the original neural network. There are additional hidden layers in the input layer and the output layer. It can significantly reduce the amount of calculation and the difficulty of analysis by way of gradient descent.

7.6.6 Convolutional Neural Network

But after the rise of the BP neural network, it can be found that the computational load of the BP neural network is still huge. It sometimes fails to give the optimal solution within our acceptable time range, or it takes too long to provide the optimal solution, which does not meet the needs of some of our applications. Then came the CNN, which is also a kind of neural network algorithm in essence, but it optimizes the content in the BP neural network, makes the calculation faster, and can get the most on many problems. It improves the efficiency of its analysis by processing related information highly concurrently. At the same time, it dramatically reduces the computational complexity between BP neural networks. Therefore, CNN can currently reach the optimal solution faster on many problems.

7.6.7 Recurrent Neural Network

Recurrent neural network models currently trained through this learning algorithm have reached a result that can write poems. It is not a single-term output, and it is to use the previous output as a re-input and re-enter the model line calculation. This model is impressive because it has particular memory ability.

7.7 TRANSFER LEARNING

In traditional machine learning, multiple models must be trained for specific problems, each designed to solve a particular ML type of problem. In transfer learning, it tries to store the results of training in a particular field (it is called the original domain) as knowledge. We train the algorithm in the original domain. After training, it can be hoped that it can solve new problems, which are called targeted tasks. After it has been put in the target task, it can infer new issues based on the knowledge deposited in solving problems in the original domain and can get results without the need for training. Transfer learning transfers the results of previous learning to new situations.

7.8 RELATION TO AI

However, an increasing emphasis on the logical, knowledge-based approach caused a rift between AI and ML. Probabilistic systems were plagued by theoretical and practical problems of data acquisition and representation. By 1980, expert systems had come to dominate AI, and statistics was out of favor. Work on symbolic/knowledge-based learning did continue within AI, leading to inductive logic programming, but the more statistical line of research was now outside the field of AI proper, in pattern recognition and information retrieval. Neural network research had been abandoned by AI and computer science around the same time. This line, too, was continued outside the AI/CS field, as "connectionism" by researchers from other disciplines, including Mr. Hopfield, Mr. Rumelhart, and Mr. Hinton. Their main success came in the mid-1980s with the reinvention of backpropagation.

7.9 GENERATIVE ADVERSARIAL NETWORK (GAN)

The word "Adversarial" in the generative adversarial network (GAN) means that there is a competitive relationship between the two networks generated. Among these two networks, one is responsible for generating samples, and the other is for determining the correctness of the pieces. The group that produces the samples hopes to fool the group that discriminates the samples. The group that discriminates the samples hopes not to be fooled by the results of the group that generates the pieces; they have a particular relationship between competition and confrontation. It is more accurate to create a learning result in such a relationship.

7.10 DATA MINING

It would be incomplete to discuss AI without the concept of its predecessor. Data mining (DM) is the process of finding previously unknown, profitable, and valuable patterns embedded in data with no prior hypothesis. It is the process of analyzing data from different perspectives, summarizing it into useful information, and finding correlations or patterns among datasets in large relational databases. The objective of DM is to use the discovered ways to help explain current behavior or to predict future outcomes. DM borrows some concepts and techniques from several long-established disciples, viz. AI, Database Technology, ML, and Statics. The field of DM has, over the past couple of decades, produced a wide variety of algorithms that enable computers to learn new relationships/knowledge from large datasets.

Although DM is a relatively new term, the technology is not. It has witnessed a considerable growth of interest over the last couple of years, which is a direct consequence of the rapid development of the information industry. Historically, DM has evolved into a mainstream technology because of two complementary yet antagonistic phenomena: the data deluge, fueled by the matured database technology and the development of advanced automated data collection tools such as the Logging While Drilling and the Measurement While Drilling; starvation for knowledge, defined as the need to filter and interpret all these massive data volumes stored in huge databases, data warehouses, and other information repositories. DM can be thought of as the logical succession to IT.

A summary of the evolution of DM over the past years is shown in Figure 7.2.

7.10.1 Relationship with Data Mining

ML and DM often employ the same methods and overlap significantly, but while ML focuses on prediction, based on known properties learned from the training data, DM focuses on the discovery of (previously) unknown properties in the data (this is the analysis step of knowledge discovery in databases). DM uses many ML methods but with different goals; on the contrary, ML also employs DM methods as "unsupervised learning" or as a preprocessing step to improve learner accuracy.

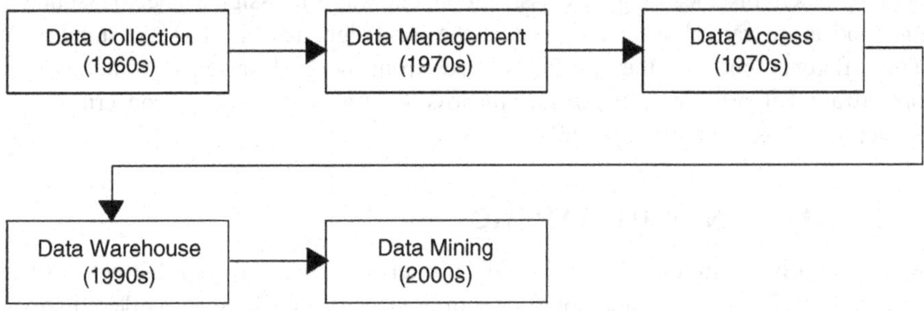

FIGURE 7.2 Flow diagram of the evolution of data mining.

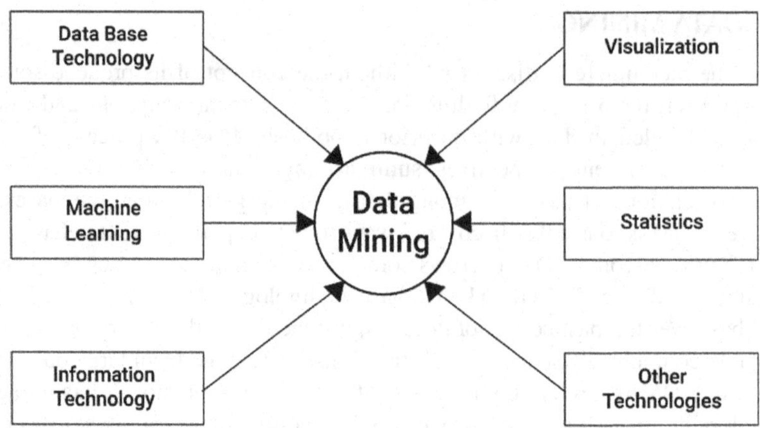

FIGURE 7.3 Relation between machine learning and data mining.

Much of the confusion between these two research communities (which do often have separate conferences and separate journals, ECML PKDD being a significant exception) comes from the basic assumptions they work with: in ML, performance is usually evaluated with respect to the ability to reproduce known knowledge, while in knowledge discovery and data mining (KDD), the critical task is the discovery of previously unknown knowledge. Evaluated with respect to general knowledge, an uninformed (unsupervised) method will easily be outperformed by other supervised methods, while in a typical KDD task, supervised methods cannot be used due to the unavailability of training data. The relationship between ML and other technologies, including DM, is explained in the block diagram shown in Figure 7.3.

7.11 RELATION TO OPTIMIZATION

ML also has intimate ties to optimization: many learning problems are formulated as minimization of some loss function on a training set of examples. Loss functions express the discrepancy between the predictions of the model being trained and the actual problem instances (e.g., in classification, one wants to assign a label to instances, and models are trained to correctly predict the pre-assigned labels of a set of standards). The difference between the two fields arises from the goal of generalization: while optimization algorithms can minimize the loss on a training set, ML is concerned with reducing the loss on unseen samples.

7.12 RELATION WITH STATISTICS

ML and statistics are closely related fields in terms of methods but distinct in their principal goal: statistics draw population inferences from a sample, while ML finds generalizable predictive patterns. According to Mr. Michael I. Jordan, the ideas of ML, from methodological principles to theoretical tools, have had a long pre-history in statistics. The term "data science" also suggests as a placeholder to call the overall field. ML is a technology that strongly overlaps with the methodology of statistics. From a historical/philosophical viewpoint, ML differs from statistics in that the focus in the ML community has been primarily on the accuracy of prediction, whereas the emphasis in statistics is typically on the interpretability of a model and/or validating a hypothesis through data collection.

7.13 SUPERVISED MACHINE LEARNING

Regression is a technique used to predict the value of a response (dependent) variable from one or more predictor (independent) variables.

The most commonly used regression techniques are linear regression and logistic regression.

The basic theory underlying these two prominent techniques includes explaining many other vital concepts like gradient descent algorithm, over-fit/under-fit, error analysis, regularization, hyper-parameters, and cross-validation techniques involved in ML.

7.13.1 Linear Regression

In linear regression problems, the goal is to predict a real-value variable y from a given pattern X. In the case of linear regression, the output is a linear function of the input. Let Ý be the output, our model predicts:

$$\dot{Y} = WX + b. \tag{7.2}$$

Here X is a vector (features of an example), W is the weight (vector of parameters) that determines how each component affects the prediction, and b is the bias term. So, the task T is to predict y from X; now it needs to measure performance P to know how well the model performs.

Now to calculate the performance of the model, first the error of each example i is calculated as:

$$ei = abs\left(\dot{Y} - Yi\right). \tag{7.3}$$

The error value has been taken into consideration, both positive and negative values of the error.

Finally, the mean for all recorded absolute errors (Average sum of all fundamental errors) should be calculated.

Mean absolute error (MAE) = Average of all absolute errors

$$MAE = 1/m \ \Sigma abs(i - Yi) \tag{7.4}$$

A more popular way of measuring model performance is using mean squared error (MSE): average of squared differences between prediction and actual observation.

$$\text{MSE} = 1/2\text{m}\,\Sigma\left(\dot{\text{Y}}\text{i} - \text{Yi}\right)2 \tag{7.5}$$

The mean is halved (1/2) as a convenience for the computation of the gradient descent (discussed later), as the derivative term of the square function will cancel out the 1/2 term. For more discussion on the MAE vs. MSE, please refer to equations 7.4 and 7.5.

To minimize the error, the model, while experiencing the examples of the training set, updates the model parameters W. These error calculations, when plotted against the W, are also called cost function J(w), since it determines the cost/penalty of the model. So minimizing the error is also called as minimization of the cost function J.

7.14 GRADIENT DESCENT ALGORITHM

In the gradient descent algorithm, we start with random model parameters, calculate the error for each learning iteration, and keep updating the model parameters to move closer to the values that result in minimum cost. Repeat the process until minimum price:

$$\text{wj} = \text{wi} - \partial/\partial\text{wi}\ \text{J}(\text{W}). \tag{7.6}$$

In the above equation, we are updating the model parameters after each iteration. The second term of the equation calculates the slope or gradient of the curve at each iteration.

The gradient of the cost function is calculated as a partial derivative of cost function J with respect to each model parameter WJ; j takes the value of the number of features (1 to n)]. α, alpha, is the learning rate, or the value that indicates how quickly we want to move toward the minimum. If α is too large, we can overshoot. If α is too small, it means small steps of learning; hence, the overall time taken by the model to observe all examples will be more. There are three ways of making gradient descent, which are discussed in the following subsections.

7.14.1 Batch Gradient Descent

It uses all of the training instances to update the model parameters in each iteration.

7.14.2 Mini-Batch Gradient Descent

Instead of using all examples, it divides the training set into smaller sizes called batch denoted by "B." Thus, a mini-batch "b" is used to update the model parameters in each iteration.

7.14.3 Stochastic Gradient Descent (SGD)

The parameters may be updated by using only a single training instance in each iteration. The training instance is usually selected randomly. SGD is often preferred to optimize cost functions when there are hundreds of thousands of training instances or more, as it will converge more quickly than batch gradient descent.

7.14.4 Logistic Regression

Logistic regression is named for the function used at the core of the method, the logistic function.

The logistic function, also called the sigmoid function, was developed by statisticians to describe properties of population growth in ecology, rising quickly and maxing out at the carrying capacity of the environment. It's an S-shaped curve that can take any real-valued number and map it into a value between 0 and 1, but never precisely at those limits:

$$1/(1+e\text{^-value}), \tag{7.7}$$

where "e" is the base of the natural logarithms (Euler's number or the EXP(e) function in your spreadsheet) and the value is the actual numerical value that you want to transform. Figure 7.4 shows a plot of the numbers between –5 and 5 transformed into the range between 0 and 1 using the logistic function.

7.14.5 Representation Used for Logistic Regression

Logistic regression uses an equation as the representation, very much like linear regression.

Input values (x) are combined linearly using weights or coefficient values (referred to as the Greek capital letter beta) to predict an output value (y). A key difference from linear regression is that the output value being modeled is a binary value (0 or 1) rather than a numeric value.

Below is an example logistic regression equation:

$$y = e\text{^}(b0+b1*x)/(1+e\text{^}(b0+b1*x)), \tag{7.8}$$

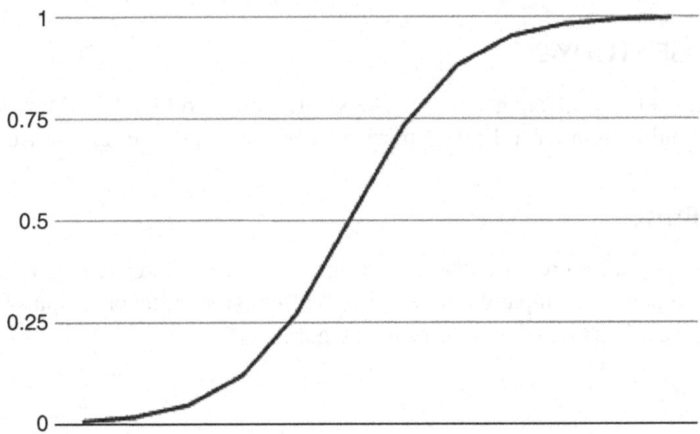

FIGURE 7.4 Logistic function.

where y is the predicted output, b0 is the bias or intercept term, and b1 is the coefficient for the single input value (x). Each column in your input data has an associated b coefficient (a constant real value) that must be learned from your training data. The actual representation of the model that you would store in memory or in a file is the coefficients in the equation (the beta value or b's).

7.15 CASE STUDY-1

Using data to drive positive business results is discussed widely, and many companies have recognized the potential of AI—and ML in particular—to implement robust solutions based on data.

In practice, companies can benefit from a wide range of machine learning applications, but it's not always obvious which solutions are the most feasible and effective for achieving the desired results. The best techniques available to solve a problem vary depending on the companies' individual goals, resources, and data. To get an idea of how ML can create business value in different industries utilizing different techniques, we came up with a summary of some projects that we have carried out at Tryolabs. Each case study touches upon the company's challenge, the custom ML system we built, the results, and the approach we chose to achieve them. All solutions follow the steps outlined below:

- Understanding the business challenge
- Evaluating different approaches to solve the challenge
- Building proofs of concepts or production-ready systems
- Evaluating results and performing additional iterations, if necessary

Some case studies consist of end-to-end solutions that we implemented from scratch and put into production, while others are proofs of concepts used by companies to establish ML roadmaps. It is hoped that the following case studies help you become more familiar with the tremendous opportunities that exist around using data to drive business results and encourage you to jump in on an ML endeavor in your organization.

7.16 CASE STUDY-2

Building an automated category tree: The solution was prepared for Mercado Libre, the largest online retailer in Latin America, headquartered in Buenos Aires.

7.16.1 Before

Inefficient system for creating item categories across 16 different category trees (one for each country the company operates in). Ineffective re-categorization of products and repetitive allocation of new items to categories.

7.16.2 After

Autonomous category creation as well as efficient product classification and proper categorization of millions of new or previously misclassified items with little to no human supervision.

7.16.3 Results

- 10× faster creation of new item categories
- Daily re-categorization of 25k new items
- 5× increase in productivity

7.16.4 Approach

Analyze massive amounts of historical data and implement a system to automatically suggest missing categories. Create a model using product titles, product descriptions, and users' navigation data to improve manual categorization efficiency. Build an intuitive and intelligent front-end allowing for human validation of the automated category suggestions so that the system can learn from feedback.

7.16.5 Technical Details

Experiment with various algorithms in order to generate new categories: K-Means, hierarchical clustering, suffix tree clustering, custom vectorization techniques, and distance functions. R&D finally leads to the creation of a custom overlapping clustering algorithm using neural networks and users' navigation data preprocessed directly from Hadoop. Implement a recursive neural network long-short term memory (LSTM) for item classification, relying on page views and item data as inputs.

7.17 CASE STUDY-3

This solution was taken from M/S Halliburton, USA, an oil service company, for repairing preventive maintenance of oil well pump.

7.17.1 Before

Data collected from sensors in oil well pumps are analyzed manually by employees in order to assess the health of the machinery. The ability to detect faults at scale is minimal.

7.17.2 After

Scholarly articles for A-environment automated the analysis of several dozen metrics that were collected from the pumps and send notifications to operators in case of warning signs. Engineers have access to a dashboard that displays the health of machinery and problematic values. The results are listed as follows:

- Capacity to analyze data from 2000+ wells in near real time
- Ability to detect new relevant alerts
- Reducing false-positive alerts by 50%

7.17.3 Approach

Build a custom front-end for petroleum engineers to tag or label regions of data of interest. Use data to train an ML model that keeps track of changes in performance metrics and predicts potentially unhealthy scenarios.

7.17.4 Technical Details

Develop a feature extractor for time series data, using NumPy, SciPy, and Pandas. Build a rules engine using time series analysis, as well as ML models, to automatically detect events and, from those events, generate different types of alerts related to the health of the wells. Implement a React application to display current metrics for the other oil well pumps and the generated signals, as well as to enable operators to give feedback on alerts. The user interface additionally allows operators to use their expert knowledge to tag healthy/unhealthy time ranges, along with other possible predefined points of interest. The labels provided by the operators are leveraged to implement an automatic feedback loop.

BIBLIOGRAPHY

Brownlee, J. "Logistic Regression for Machine Learning." Machine Learning Mastery. 2016 April 1. Updated, 2023 December. https://machinelearningmastery.com/logistic-regression-for-machine-learning/.

Kumar, Aviral, Aurick Zhou, George Tucker, and Sergey Levine, "Conservative Q-Learning for Offline Reinforcement Learning." *Advances in Neural Information Processing Systems 33 (Neur IPS 2020)*, 2020. 1–13.

Penn, Joe, *Machine Learning with Python: The Step-By-Step Practical Guide to Grasp Machine Learning, Build Algorithms with Python and Become a Data Scientist*, Amazon. 2023 November. 45–63.

Raschka, Sebastian, Yuxi (Hayden) Liu, and Vahid Mirjalili. *Machine Learning with PyTorch and Scikit-Learn*, Packt, Mumbai. 2022 February. 14–19.

Wikipedia. "Arthur Samuel. (Computer Scientist)." n.d. https://en.wikipedia.org/wiki/Arthur_Samuel_(computer_scientist).

8 Intelligent Drilling and Well Completion

8.1 INTRODUCTION

Technological innovation in drilling is a good strategy that will affect the drilling space positively due to the application of artificial intelligence (AI) with predictive analysis. The industry will be able to automate the drilling process to mitigate some of the important drilling dysfunctions such as differential sticking, stuck pipe, excessive downhole vibration, managed pressure drilling, hole cleaning, equivalent circulation density, and pressure transients. These downhole phenomena can be addressed by an input of good reliable data curated from well site records, including borehole equipment and run data, analyzed offset data, survey data, geological data, and the well plan. AI can address these issues by using an analyzing downhole data acquisition system (sensor data). Potential anomalies can be flagged promptly using reliable indicators. The end result of an integrated automated platform in the digital drilling is optimal control of operation parameters, which directly affect the rate of penetration (ROP). Another key benefit of AI-infused drilling is safety; fewer people will be involved in the process, and the skill set requirement will be increased. Numerous drilling operations can be changed or improved using computational intelligence and mathematical optimization, including the following:

- Reduced drill pipe wear
- Bit directional guidance
- Directional drilling decision automation
- Measurement while drilling (MWD)
- Managed pressure drilling
- Mud programs
- Casing while drilling
- Rig operation process
- Increasing production with targeted precision
- Decreasing drilling time
- Controlling net productive time (NPT)
- Well site geology automated formation top correction

The benefits of AI are provided as follows:

- Detection of cyber threats and rig control system security lapses.
- Prescriptive maintenance.
- Open window for performance optimization—offset data analyzed from bottom-hole sensors will be used to make real-time, closed-looped predictions that will help intelligent node systems react and resolve current drilling dysfunctions.

DOI: 10.1201/9781003307723-8

- Better Safety: This is one of the key issues in drilling industry, and the ability to change the status quo and take off the drilling rig will be a win for everybody. Drillers will be able to monitor multiple wells from remote locations while focusing on monitoring drilling performance/challenges, which will cut the cost of drilling significantly. AI will help curb the risk and exposure to unsafe working conditions for many employees.
- More collaboration among service providers—with an open platform, a correlating increase will occur in software innovation platforms for performance-enhancement applications, which also promotes positive problem-solving competition.
- Intelligent drilling automation will not change the way the industry drills wells but rather will take current drilling practices and improve real-time decision-making.

Intelligent completions incorporate permanent downhole sensors and surface-controlled downhole flow control valves, enabling an engineer to monitor, evaluate, and actively manage production (or injection) in real time without any well interventions. Well completion engineering is a relatively independent work linking drilling engineering and production engineering, which combines engineering, geology, and exploration. Many major oil and gas fields have gradually entered into the deepwater development stage, and many unconventional reservoirs like shale gas also launched large-scale development. Under this, horizontal well completion has received great attention and achieved rapid development. In order to develop hydrocarbon resources to the maximum extent, many advanced horizontal well completion tools and instruments have been developed successively, forming a series of new completion technologies, such as sand control completion for high-end horizontal wells, multistage fracturing completion, self-adaptive inflow control, and intelligent well completion (IWC). This field has developed into one independent subject. High attention is needed for horizontal well completion technology to satisfy oil and gas production demands, protect the reservoir, and prolong the life of oil and gas wells as far as possible. Selecting right and suitable completion mode and completion parameters directly determines the future development and production plan. The well completion practice shows that the completion method and the matching completion technology, which are suitable for the geological conditions and fluid properties of the oil and gas field, cannot only significantly improve the production capacity of oil and gas wells and efficiently develop oil and gas fields but also reduce the direct oil and gas production expenditure, prolong the life of oil and gas wells, and use low investment to achieve more production. To select an appropriate well completion system, sufficient information about the petro-physical properties of reservoir rocks, properties of formation fluids, future production technologies to be adapted shall be learnt, and the advanced well completion system design software shall be applied to carry out comprehensive optimization design and propose a suitable and efficient well completion method. Selection and design of completion methods are the core link in the process of well construction since reservoir, geology, drilling, and oil recovery should all be designed and implemented based on the requirements of completion.

Analytic software provides visual views of data to help and turn them into actionable information. Machine learning software can take into account a number of factors important to overall drilling strategy. These include equipment ratings, seismic vibrations, strata permeability, and geothermal gradients. When layered, these data can be used to determine not only the optimal direction of the drill bit but also how it should be controlled (i.e., ROP) as it bores through the ground. Information is logged, contextualized, and visualized via human–machine interfaces, allowing personnel to monitor the overall performance of wells and make more intelligent decisions aimed at improving operations.

Predictive software can also be used to analyze data to determine, if downhole conditions are conducive to potentially catastrophic events such as a lost circulation, stuck pipe, or blowouts. By leveraging data to understand what contributes to the likelihood of such an event, algorithms can provide recommendations to the control system and operating personnel to minimize the odds of such an occurrence.

The purpose of development of oil and gas field required rigorous analysis of reservoir behavior to increase the production from exiting field, i.e., to increase the recovery factor. More data, such as pressure, temperature, and online water cut values, are needed to fine-tune the existing reservoir monitoring to improve these exercises. Existing wells are to be equipped smart wells with proper well completion equipment (bottom-hole pressure transmitter, temperature transmitter, flow meter, etc.). In digital oil field, all these data from individual wells are connected to the central control room, where all these data are being monitored continuously and analyzed. In offshore, an individual group of cluster wells is tested regularly through three-phase separator (multiphase meter) in order to know the fluids behavior and productivity. A block diagram of digital real-time monitoring system of oil and gas drilling, production/operation, processing, gas flaring, reservoir monitoring, pipeline transportation system, and product storage is shown in Figure 8.1, which explains the linking of field-level data to centralized data monitoring system.

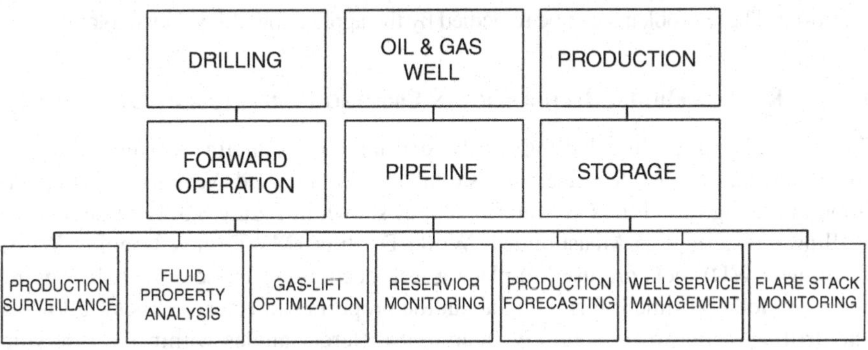

FIGURE 8.1 Diagram of digital oil field control and real-time monitoring system.

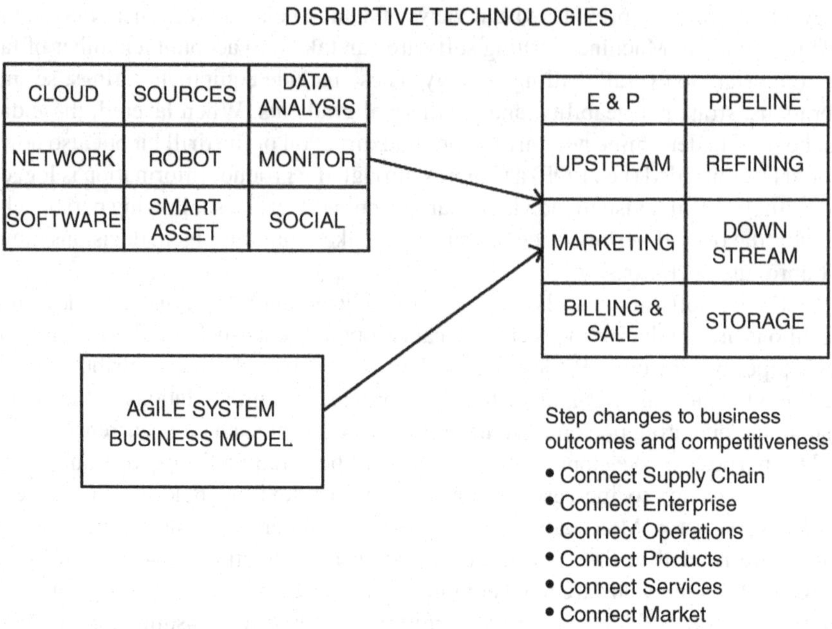

DISRUPTIVE TECHNOLOGIES

CLOUD	SOURCES	DATA ANALYSIS
NETWORK	ROBOT	MONITOR
SOFTWARE	SMART ASSET	SOCIAL

AGILE SYSTEM BUSINESS MODEL

E & P	PIPELINE
UPSTREAM	REFINING
MARKETING	DOWN STREAM
BILLING & SALE	STORAGE

Step changes to business outcomes and competitiveness

- Connect Supply Chain
- Connect Enterprise
- Connect Operations
- Connect Products
- Connect Services
- Connect Market

FIGURE 8.2 Digital transformation of an oil and gas industry.

The concept of a digital oil and gas operation can be related to supply chain, engineering consultant, service provider, workers, and consumers. A step change to business outcomes and competitiveness toward better managed system are explained in Figure 8.2.

Disruptive, old business model and work process have to be modified to present business model such as data transmission from main source to respective data acquisition center.

Numerous oil fields worldwide have been producing at rates considerably below potential values (Productivity Index). Marginal reserves have been overlooked and discarded because the technology required for profitable exploitation has been elusive, expensive, or unproven. These problems can be remedied by the application of IWC technology.

8.1.1 Role of Digital Technology in Drilling Planning and Operation

The role of logging while drilling (LWD), drilling fluid, sampling, coring, rig instrumentation, and downhole equipment is totally interconnected. Various tools in operation in the digital oil field are interlinked, as shown in Figure 8.3. New directional drilling tools, such as Measurement While Drilling (MWD) and Logging While Drilling (LWD) tools are used to place a well accurately. These downhole logging tools provide real-time information about the properties of reservoir rocks, allowing the drilling string to be precisely steered at different angles within the reservoir. These tools accurately steered the drilling string at various angles in the reservoir. This rotary steering system (RSS) also cuts formation cores ample for carrying out petro-physical study for the evaluation of reservoir rocks properties. LWD equipment saves time for the identification of producing horizon without pulling out from

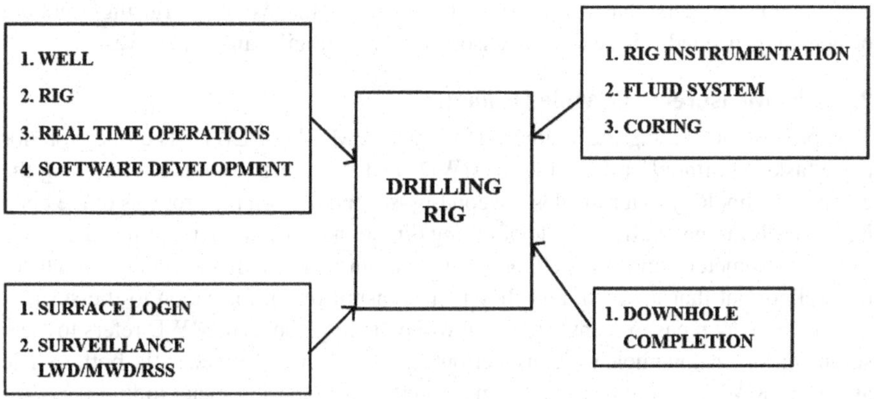

FIGURE 8.3 Workflow chart of digital oil field in drilling domain.

the hole. Both MWD and LWD tools transmit bottom-hole data to surface unit from bottom-hole sensors. Modern offshore drilling rigs are equipped with rig instrumentation system. All online data are being analyzed at drillers' cabin continuously, which is equipped with computer control joystick technology.

8.1.2 Digital Drilling Application

Recent advancements in drilling techniques have delivered new levels of operational efficiency. Yet, operators continue to strive for even higher levels of drilling performance, with many requesting drilling service providers supply higher levels of automation on their rigs. There is little question about the value of automation—other industries have proven the transformational impact that automation can have on both operations and business models. However, the following question remains: How will the drilling industry realize the full potential of automation?

In a market dominated by day-rate contracts and eroding margins, many drilling contractors struggle with finding ways to profitably supply the automation technologies requested by operators. However, the right approach can deliver operational efficiencies that are more than required to offset the investment. Efficiencies are only the most immediate tangible value delivered; other benefits include predictable, consistent results aligned with standard operational procedures and performance standards, and increased safety.

8.1.2.1 Directional Drilling

In the electronic measurement system during drilling directional well, almost every electronics-enabled MWD system has a micro-electro-mechanical system (MEMS) sensor of some type onboard. MEMS sensors have been a natural fit for sensing and quantifying secondary parameters such as drilling dynamics and dysfunction. Sensors for measuring downhole shock, vibration, and rotation parameters are readily available and fill the role well. However, MWD systems have clung to legacy sensor technology for performing critical borehole orientation measurements. An alternative to the standard orientation sensors composed of a set of triaxial (think three dimensions) inertial (mechanical, quartz-based) accelerometers and a set of triaxial fluxgate (wound-core) magnetometers that are used as the primary means of measuring wellbore placement is needed. These legacy sensors are large, fragile, and expensive.

8.1.2.2 Measurement While Drilling

This process of guiding the wellbore is known as MWD or LWD. The operators for these tasks, commonly referred to as MWDs or LWDs, are responsible for using the complex technology that provides a second-by-second feed on the progress of the bore. MWD tools are generally capable of taking directional surveys in real time. The tools use accelerometers and magnetometers to measure the inclination and azimuth of the wellbore at that location, and they then transmit that information to the surface. Although many measurements are taken while drilling, the term MWD refers to measurements taken downhole with an electromechanical device located in the bottom-hole assembly. MWD uses gyroscope, magnetometers, and accelerometer to find the borehole inclination, azimuth during drilling, and data to surface. A method of transmitting LWD and MWD data acquired downhole to the surface is using pressure pulses in the mud system. The measurements are usually converted into an amplitude- or frequency-modulated pattern of mud pulses. Well delivery in drilling planning is explained using MWD sketch as shown in Figure 8.4, for the placement of well at desired location.

8.1.2.3 Logging While Drilling

LWD or MWD (the terms are used interchangeably) is a general term to describe systems and techniques for gathering downhole data while drilling without the requirement to

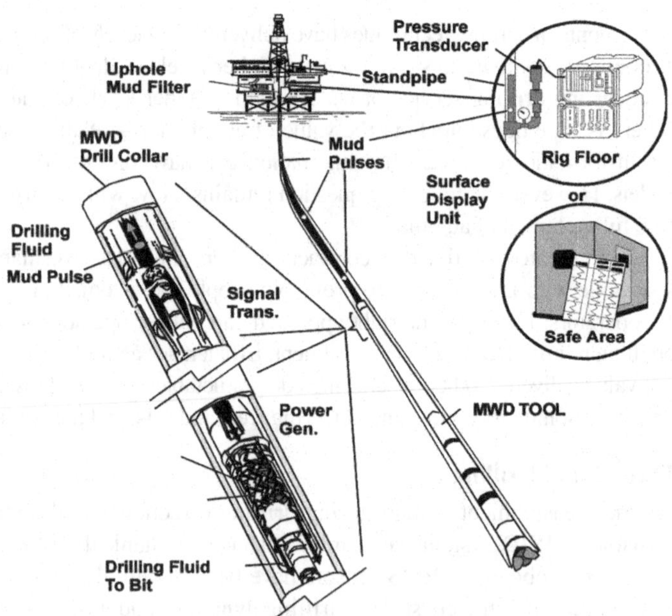

FIGURE 8.4 Detailed drawing of MWD tool for directional drilling.

remove drill pipe from the well. Providing information on porosity, resistivity, acoustic waveform, hole direction, and weight on bit, LWD transmits logging measurements at regular intervals while drilling is taking place. Data are transmitted to the surface through pulses through the mud column (also known as mud pulse of mud telemetry) in real time. LWD tools are tools that have sensors to record information about rock formation during drilling. LWD tools use sensors placed just above the drill bit to detect natural gamma rays and other phenomena as drilling progresses. LWD tools measure formation speeds while the borehole is drilled, increase ROP, improve wellbore stability and hole quality, and optimize well placement and reservoir exposure for maximum production. LWD technology acquires high-quality data for accurate geo-steering and more-informed formation evaluation, so it can work proactively during drilling.

The main difference between LWD and MWD is that whereas LWD data are recorded in memory and downloaded when the tools reach the surface, MWD data are transmitted up the drill pipe by means of a pressure wave (mud pulsing) at 3 bits/sec and monitored in real time (Figure 8.5).

8.1.2.4 Rotary Steering System

An RSS is a form of drilling technology used in directional drilling. It employs the use of specialized downhole equipment to replace conventional directional tools such as mud motors. The advantages of this technology are many for both main groups of users: geoscientists and drillers. Value and direction of steering force are first determined based on the current actual point and the expected point on the ground. The steering force is transmitted downhole. Based on steering force, each pad force is next determined in accordance with predetermined control algorithms downhole. Lastly, the pads are pushed out by applying hydraulic pressure, and the expected steering force and well trajectory are realized. RSS enables drill string rotation while deviating the wellbore. The difference in this steering mechanism makes RSS drills faster than the conventional mud motor while building or dropping well trajectory angle. But in build/drop section, ROP of RSS was higher four times than that of the conventional mud motor. The PowerDrive vortex RSS enables drilling with increased ROP, maximizes power at the bit to drill longer runs, and increases the rig's operating envelope. Its rotation conditions the hole while drilling and is proven to reduce the risk of sticking, casing wear, and drill string fatigue The rotary steering tool works in a bore hole either by pushing or holding of pads. Various tools, like pragmatic general multicast (PGM), MWD, LWD, and RSS assembly, inside the bore with respect to rotary steering drilling assembly are shown in Figure 8.6.

8.1.3 Service Provider's Dilemma

When it comes to offering automation, drilling contractors can take several approaches. One way is to combine point solutions from various providers. Each point solution addresses a specific aspect of the drilling process, such as stick/slip mitigation, directional guidance, automated sliding, or on-bottom drilling parameter management, aiding the driller in the completion of a task. Point solutions deliver value to discrete operations but are limited in the value they offer due to the need for the driller to integrate and manage them. To transform drilling operations, an automation solution needs a wider scope.

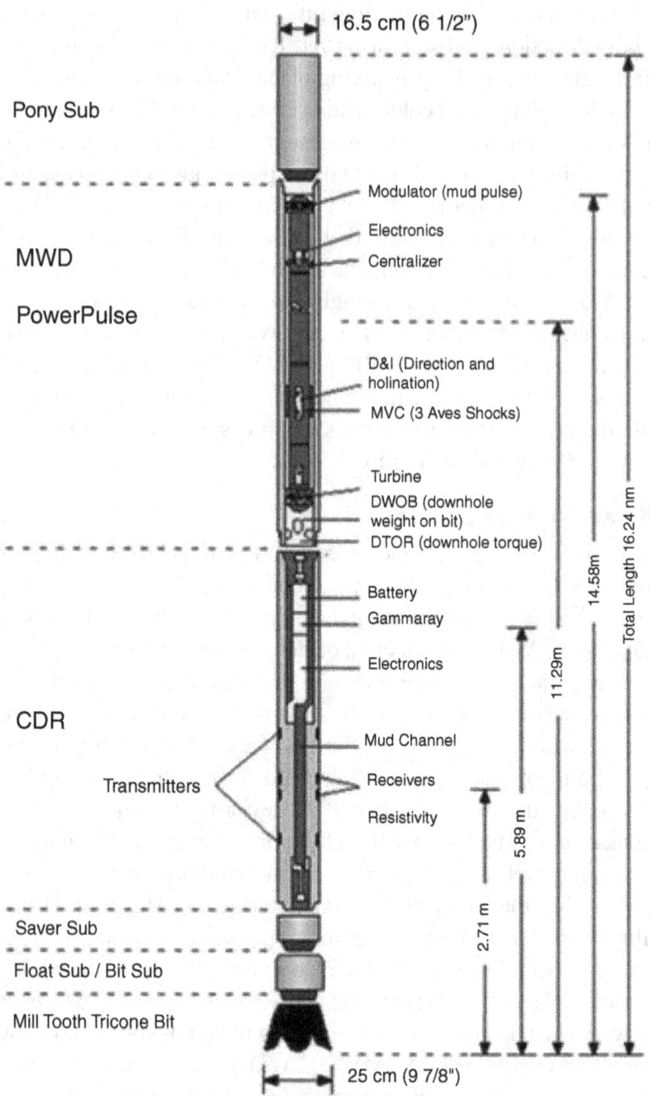

FIGURE 8.5 MWD drilling tool assembly.

An alternative to point solutions is a process control automation system that helps the driller through post-connection, on-bottom drilling, and pre-connection drilling activities. These systems are available from several vendors and include varying degrees of operational flexibility and support for dynamic drilling conditions. Automation systems augment the capabilities provided by a rig's control system. Because there is no industry-standard control system interface, a rig can only choose an automation solution that the control system vendor has specifically enabled. The well construction workflow is a continuous cycle from well planning through drilling operations, where learnings are incorporated through each cycle.

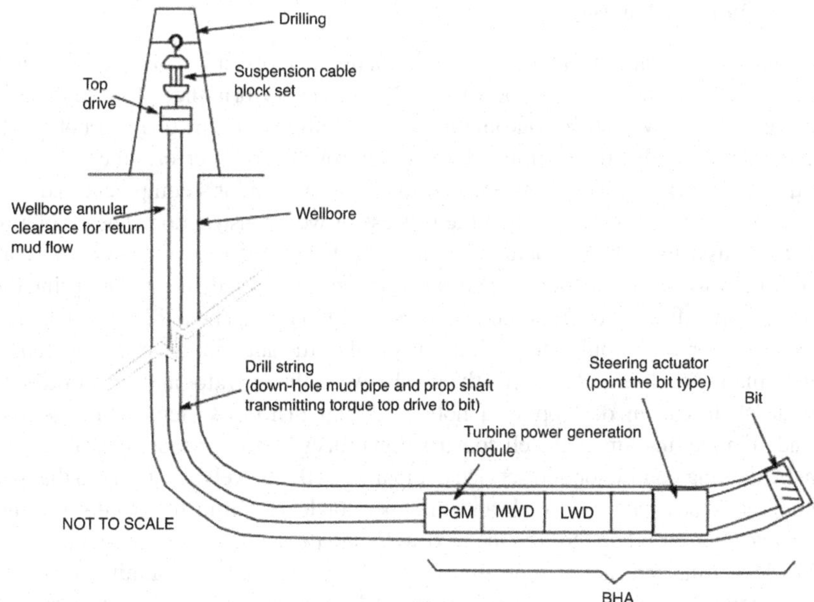

FIGURE 8.6 Rotary steering tool assembly.

Some drilling contractors have developed bespoke automation solutions for their own fleets. Automation development—which requires capabilities related to software, firmware, controls, signal processing, AI, IT security, data management, telecommunications, and configuration management—is not typically a core competency of a drilling contractor. This approach requires a significant, prolonged investment to develop and support the technology, and this investment is not guaranteed to result in positive returns. An alternative is for the drilling contractor to work with an automation supplier, who can bear the development costs, bring experience, and expertise beyond a single fleet of rigs, and amortize the investment over a larger number of rigs.

Beyond the outlay to develop the automation technology itself, there is an investment in each rig to utilize this advanced technology. Each rig's hardware, software, and firmware must be modified to host the technology. Each rig crew has to transition from manual operations to automation-assisted operations. Continually driving toward ever higher levels of performance requires continuous investment in the base technology, as well as its optimal usage.

Drilling is fundamentally a team sport where the operator, drilling contractor, and multiple service companies must work together toward a singular goal. After the largest market shock in history, all oil and gas players are under real stress. They will have to innovate out of this with technology to deliver operational efficiency, service delivery cost reduction, and capital efficiency.

Tackling the problem from only one perspective will not lead to a solution. The industry must find a way to drive performance by incorporating the technical and commercial perspectives of the contractor, service companies, and operator. What the industry needs is the integration of the digital and physical, surface operations, and subsurface knowledge, to deliver the next generation in efficiency, safety, and performance.

8.1.4 Digital Drilling

Most automation solutions focus on the automation and operation of rig equipment to achieve specific tasks. They support the driller in the performance of those tasks but miss the wider view. Drilling automation needs to align with the goals of not just the rig contractor but also the operator if it is to improve overall performance.

Equipment-oriented solutions are tied to a specific vendor's equipment and focus on equipment operation rather than the process of well construction. These solutions rely on the user to map the operator's well construction plan onto the system's capabilities. In most instances, the user does this by consulting a drilling plan, printed on paper, that specifies the operational activities at a high level. Then, they translate those into settings for the automation system. This makes the automation system dependent upon human expertise and sustains the gap between the operator and rig workflow.

Industry has been developing a holistic digital drilling system. This system is grounded in the fundamental well construction process—an operator plans a well, and then the drilling contractor and service companies drill the well according to that plan. Operational results are recorded, and the operator learns from the results obtained. Those learnings are then fed into the next well, and the loop starts over again.

While drilling contractors typically do not take part in the well planning workflows, it is essential that their rigs integrate seamlessly with the results of their customer's workflows so that they can deliver the well according to their customer's specification. Using the DrillOps on-target well delivery solution, the user starts automation with a digital plan, leaving no room for miscommunications or misunderstandings and removing the need for the user to translate the plan into settings for the automation system.

This solution creates a collaboration space around operations that encompasses office roles and rig operations. This automation can be deployed on any AC variable frequency drive rig through a rig control system-specific interface. The operator can select other solution elements that it would like to utilize, and they all work together seamlessly. The entire well delivery solution is open through extensibility mechanisms that allow drilling contractors, operators, and software vendors to build and deploy their own proprietary capabilities and leverage the solution's foundation.

DrillOps Automate, a module in the on-target well delivery solution, can drill a stand by going on and coming off bottom, following parameters set by the user or specified in the digital drilling plan. Once on bottom, the solution optimizes the ROP within the limits specified for the section in the well plan. This solution monitors and automatically mitigates dysfunctions, such as shock and vibration or hard stringers.

It is important to realize that, just as self-driving cars are not merely the collection of driver assistance systems, such as adaptive cruise control and blind spot monitoring, this module is not simply a collection of reflexive responses but, like a self-driving car, is designed to adapt and respond appropriately to ever-changing conditions with minimal user intervention.

8.1.5 Continuous Improvement and Sustainable Performance

Schlumberger has deployed the automation solution in various rig environments, both on land and offshore, and each deployment has brought new lessons. Early proof-of-concepts were deployed on rigs in North America, the Middle East, and

Mexico, and they connected real-time analysis desktop software applications to the rig's control system.

These proof-of-concepts automatically adjusted control system parameters to optimize ROP and demonstrated the value of automation. However, it became clear that a scalable solution would require an industrialized automation platform designed for system stability, streamlined deployment, and safety for real-time process control.

It also became apparent that an automation solution cannot execute as a disconnected system on a remote rig but must be coupled with cloud capabilities for remote workflow participation, analysis, and support.

Safety is a key consideration for drilling and, therefore, a requirement for a drilling automation solution. The solution limits drilling operations to a prescribed safe operating envelope and takes advantage of the safety features and machine control of the rig control system. Drilling plan with safety is explained in Figure 8.5. Additionally, in this digital era, cybersecurity is an essential safety element that cannot be bolted on to a system.

Well construction operations have procedures associated with them. Unfortunately, these procedures are not always followed. An early focus for deployments in the Middle East and North America was to drive procedural adherence. Graphs of real-time data show that automation can adhere to procedures in the way that no human ever could.

The block speed graph profiles for a series of connections in automation and manual modes and illustrates the difference in procedural adherence achievable through automation. With procedural adherence comes consistency, without which it is impossible to learn or improve as there is too much variability.

Once consistency has been achieved and the right procedures have been chosen, they can be deployed to other rigs or projects to enforce standardization. Analysis of the post-connection procedures on two rigs in the Middle East, prior to deployment, showed that there were four methods used for post-connections—one for each driller per shift per rig. After rolling out automation, the same procedures were applied with similar results across all rigs. Replicating these results using the traditional training and dashboard monitoring approach is extremely difficult, especially when factoring in the rotation of individuals on the drilling crews.

Standardized procedures have delivered an important benefit in reducing NPT. By enforcing and standardizing the same procedure throughout a fleet of rigs, all rigs had the same approach to detect and manage hard stringers, drilling dysfunctions, over pulls, and other types of drilling challenges. Once the stick/slip condition is mitigated, the system can resume ROP optimization. After rolling out the DrillOps solution, tool failure NPT was reduced by 90%.

With each iteration through the plan, drill, record, and learn well construction cycle, the optimal performance level comes closer within reach because it is built on a strong foundation supporting the overall process.

It was a great effort to implement the intelligent transmitter technology in oil and gas well in drilling and well completion in designing and operation. Oil well drilling technology is discussed in this chapter including manpower, management while drilling, LWD, etc., which will be interconnected with iCloud technology facilitates tight-knit collaboration and knowledge sharing among cross-functional teams, bringing the full weight of enterprise knowledge and experience to increase drive efficiency.

As the industry's only complete and fully integrated technology system, Digital Well Program will help drilling/well completion engineer to accelerate end-to-end well delivery with preplanned drilling/well completion.

8.1.6 New High-Tech Drilling Planning

Concepts, Collaboration, and Right-time Decisions delivers to the reader a roadmap through the fast-paced changes in the digital oil field landscape of technology in the form of new sensors, well mechanics such as downhole valves, data analytics and models for dealing with a barrage of data, and changes in the way professionals collaborate on decisions. Drilling planning, design, and optimization of operation flow charts are generated before the start of the operation. This network explained how complex nature of coordination with various phases needed while implementation of digital technology.

8.2 WELL COMPLETION

Well completion is the process of making a well ready for production (or injection) after drilling operations. After a well has been drilled, the drilling fluids need to be displaced from the hole, and the well would eventually close in upon itself. Well completion equipment is used to bring an oil well to the completion state so that first the test production and later on the commercial production can be started from it. Well completion incorporates the steps taken to transform a drilled well into a producing one. These steps include lowering production casing, cementing, hermetical testing of the casing, perforation of prospective zone, installing a production tubing, and production x-mass tree. Casing ensures that this will not happen while also protecting the well stream from outside incumbents, like water or sand.

Completion fluids help to improve well productivity by reducing damage to the producing zone, and they can help to prepare, repair, clean out, and complete the wellbore during the completion phase. Completion fluid is a solids-free liquid used to "complete" an oil or gas well.

8.2.1 Perforations

Perforations are an elemental piece of the inflow section of the well and have a significant impact on the total completion efficiency. This chapter describes the methods of creating the best flow path for a particular completion. It also contains information on completion diagnostics and candidate selection for situations in which re-perforating could improve production. The intent of this chapter is to familiarize the engineer with methods and techniques to improve the flow path, not all of which involve perforating equipment.

The primary objective of a perforating gun is to provide effective flow communication between a cased wellbore and a productive reservoir. To achieve this, the perforating gun "punches" a pattern of perforations through the casing and cement sheath and into the productive formation. The most common phasing values of perforating guns are 0°, 180°, 120°, 90°, and 60°. Several specialty guns, offering higher density charge application and guns for sand control or casing protection, may offer phasings that increase the linear distance between the charges in a direct line along the gun body. Optimizing petroleum production is an exercise in removing pressure drops in a flowing system that stretches from the outer boundaries of the reservoir to the sales line. The perforating process is one element in this engineering exercise. To optimize the whole process, the most severe pressure drops must be examined and removed. As each pressure drop is reduced, the increased flow may change the requirements in another cross-section of the well. Increasing the flow capacity of the reservoir by stimulation or flooding places a greater capacity requirement on the perforations (Figure 8.7). Other well completion actions, such as gravel packing, change the flow requirement on the perforation by filling the perforation with gravel. Each action changes the criteria for perforation design; therefore, initial perforating designs may not be optimal for later well production. Well design should allow for flexibility in the completion type, which allows for adding perforation density in a zone or perforating other zones after the well has been evaluated or produced.

Phasing is the angle between two charges that shows a common perforator phasing. Although there are many possible angles, the five common values are 0°, 180°, 120°, 90°, and 60°. 0° phasing aligns all the shots in a row. The gun should be decentralized, typically against the low side of the casing, so that performance from small charges is maximized by minimizing the clearance between the gun and the casing wall. 0° phasing normally is used only in the smaller outside diameter guns or guns in very large casing. 0° phasing has some drawbacks because putting all the shots in a row lowers tubular yield strength, makes the casing more susceptible to splits, and collapses at shot densities greater than 6 SPF. Fracture stimulating in wells that were perforated with 0° phasing may result in a slightly higher incidence of fracturing screen outs than with 60°, 90°, or 120° phasing. It is unknown whether the screen outs result from the smaller entrance holes or from one wing of the fracture wrapping around the pipe.

Well clean-up is an activity performed on producing wells where hydrocarbons and other fluids are brought to the surface through permanently installed well completion equipment in order to clean the well and perforations before the well is put back into production.

Smart well completion is one of these intelligent or modern techniques that include permanent downhole sensors and surface-controlled downhole flow control valves, allowing to record, evaluate, and actively manage production in real time without any well interventions.

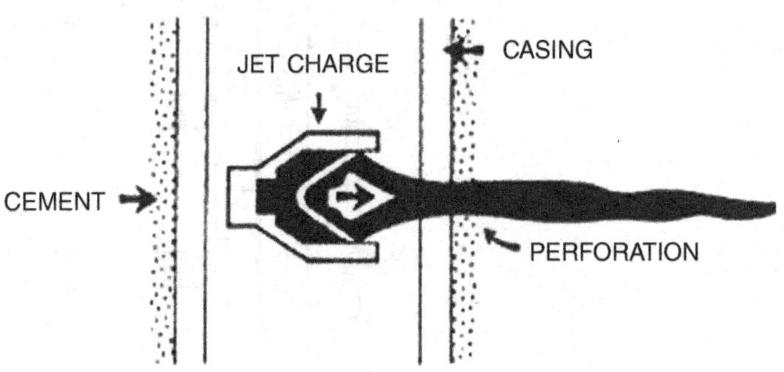

FIGURE 8.7 Typical formation fluid flow through perforation.

It is usually a packer run in close to the bottom end of the production tubing and set at a point above the top of perforations or sand screens. In wells with multiple reservoir zones, packers are used to isolate the perforations for each zone, and each zone flows through sliding sleeves or side pocket mandrel or Inflow Control Valve (ICV).

8.2.2 Definition and Principle of Intelligent Well Technology

Intelligent Well Technology, or Intelligent Completion, is a complete system of the production well that enables continuous and real-time reservoir management. Therefore, the data such as downhole temperature, pressure, flow, and composition collected by the well sensor are fed back to the up-hole system in real time. A "smart well" is a well that is equipped with pre-installed high-tech devices in the wellbore that help in monitoring and controlling the well from the surface electrically or hydraulically. They would also have the ability to self-control, collect, and analyze the data. The core of the technology is to form a closed-loop control; therefore, the data such as downhole temperature, pressure, flow, and composition collected by the wellbore sensor are fed back to the up-hole system in real-time. And then, the same data will be deeply processed, analyzed, and judged on the software platform. After that, a reservoir management decision instruction is formed and transmitted to the downhole production tool for remote operation via the wireless communication control system.

A basic sketch of an IWC is explained in Figure 8.8, which is equipped with inflow control device (ICD) and packer. Intelligent Well Technology helps to connect the surface production system with the downhole sensors. Its specific application in reservoir development is mainly to optimize production and control the

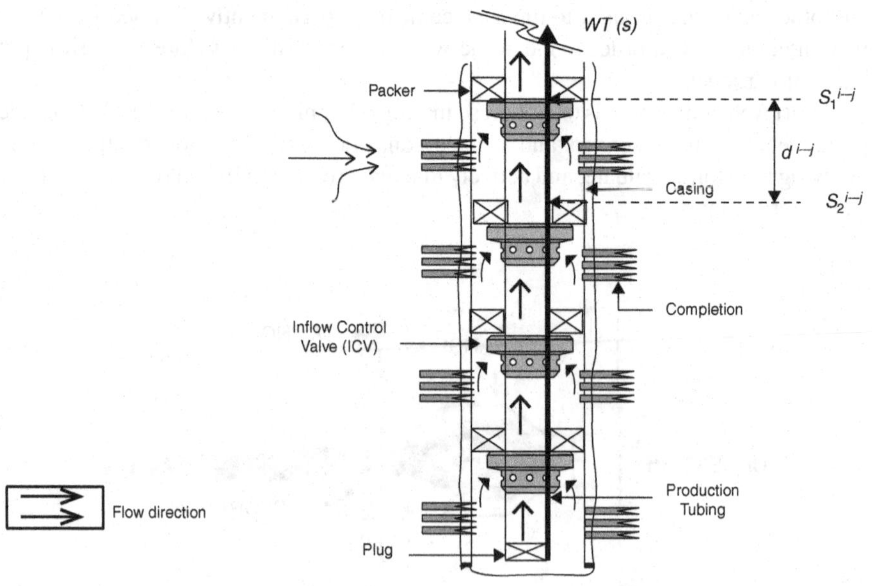

FIGURE 8.8 Intelligent well completion.

occurrence with the goal of maximizing recovery. Faced with more and more complex reservoir production projects, multi-layer combined production can give full play to the production capacity of oil well, which is better than sequential production.

One of the applications of intelligent well technology is to shut down the produced water that affects the production handling system. When the pressure is not balanced, the intelligent downhole throttling technology and time balance can be adopted. The complete intelligent well needs to have factors such as interlayer isolation, flow control, mechanical oil recovery permanent monitoring, and sand control taken into account. In this way, engineers can monitor and control the oil and gas production of single-well multi-segment oil and gas production or single-branch wellbore in multi-branched well in real time. The resultant real-time reconfiguration of the well structure will help to improve well production. In addition to that, the number of drilled wells is less with fishbone technology because of the larger drainage area achieved from one fishbone well. Another main advantage is ensuring the right placement of the fracture, to assure fracture opening and increase the efficiency of fluid transportation from the reservoir to the wellbore, unlike multi and hydraulic fracture methods. That would lead to a delay in the decline of the productivity of the well. Moreover, fishbone operations are considered more efficient compared to multi - and hydraulic fracture methods when it comes to operation time and overall project economics. However, pre-job designing has to be precise in obtaining the number, length, and angles of the ribs to optimize production from the reservoir and to avoid accessing unwanted formations/features. That would be obtained from understanding the reservoir characterizations and using simulation software to predict the optimal fishbone design. Fishbone can be considered as a type of commingle production since it can access different zones, especially in compartmentalized reservoirs.

8.2.3 An Intelligent Well Completion System

8.2.3.1 The Simulation Methodology

The objectives of simulation of this type of IWC include the following:

1. To carry out extensive study on IWCs and their applications.
2. Apply a reactive "trial-and-error" method of IWC, by modifying drilling, completion, production, and reservoir parameters to obtain maximum hydrocarbon recovery and minimal water production (<90% water cut).
3. Analyze and compare results by simulating the modifications in Objective 2 with a numerical simulation package (Eclipse, Schlumberger software) to predict production.
4. Employ ICDs for IWC applications.
5. Carry out reliability and risk register surveys for optimal IWC.
6. Create an algorithm for Proactive IWC.

The conceptual model enables the separate phases to flow with different velocity in the reservoir. Flow through the ICV (modeled as a constriction with a specified cross-sectional area) imposes an additional pressure drop. The condition that triggers this action and, the action itself has to be specified by the action keyword in Eclipse software. The action chosen is based on the production profile of a particular segment (representing flow from a particular zone). The time (e.g., each simulator time step, in this case 20 years in 240 time steps) at which the simulator checks whether the condition is exceeded was also specified. ICV action is modeled by introducing an extra pressure drop (calculated from the reduced choke dimensions with the sub-critical choke model) into the ICV segment when certain conditions are met. In some case, the segment with the highest Segment Water Flow Rate is commanded to shut off after every time step. The actions of the ICVs are then considered under varying reservoir conditions, namely, varying permeability, wettability, and varying skin factors.

ICDs (Figure 8.8) or Inflow Control Valves (ICV) are considered as key equipment in IIWC. This provides variability in the control/regulation of fluid flow through the single or multilateral wells, by utilizing multiple open and close flow ports. ICD are usually of two types: binary (open and close control) and variable (intermittent, stage-wise control).

IWC technology is applicable to both conventional and unconventional hydrocarbon wells. The technology is being considered in the completion of heavy oil fields and is highly beneficial in Enhanced Oil Recovery operations. The goal of IWC is the automation of as much of the production process as achievable.

8.2.4 Design Plan for the Development of Intelligent Well Technology

The key technologies of intelligent well have achieved some success in India, but these achievements only have some functions of intelligent well system. And there is still a big gap compared with the mature and complete intelligent well system abroad. The British Shell Group has summarized the research direction of intelligent well technology in the future:

a. Production control of multiple reservoirs can be realized by a single well.
b. Intelligent well technology minimizes the number of workovers in daily production.
c. Rather than just the optimization of the basic production unit, the whole oil field production system should be optimized from a systematic point of view, so as to achieve excellent compatibility between downhole tools and ground equipment.
d. Automatic production of oil well and optimization of production strategy of oil well.
e. The continuous service life of the intelligent well system after putting into use should be 10 years, and the reliability should be more than 95%.

The major assumptions for the development of intelligent well technology in India are as follows:

The intelligent wells offer many advantages over conventional completions. They primarily range from accelerated production, increased recovery, reduced or eliminated intervention, and improved reservoir performance. A practical solution to minimizing this problem is to increase oil production and recovery.

Intelligent completion has significantly improved the recovery factor from new and existing oil and gas fields. The primary goal is to simulate the application of IWC (Intelligent Well Completion) technology using the Schlumberger Eclipse simulator. To achieve this objective, the well design of Field "X" is thoroughly considered from the drilling to the completion phases and optimized at each stage. The simulation includes using Intelligent Well Inflow Control Devices (ICDs) and downhole sensors. The ICD offers the ability to open and close sections of the borehole, while the downhole sensors monitor the borehole and reservoir properties, providing better reservoir management and early anomaly detection. Based on the analysis of the simulation results, risk management, sensitivity analysis, and other procedures, it was noted that the IWC technology presented an increase in oil production by approximately 50%, a higher return on investment, low attached risks within the as low as reasonably practicable (ALARP) region, a high reliability, and a minimal water production (water cut < 90%) among other added values, thus proving the multiple benefits of IWC adaptation. Although inherent challenges still need to be overcome, with ample room for growth and development, the IWC technology has the potential to alleviate the crippling challenge of the shortage of hydrocarbon energy sources.

8.3 INFLOW CONTROL DEVICES

Two main types of ICDs are used in industry: passive ICD and autonomous ICD. The passive ICDs use simple geometries to create the pressure drop, with basic losses. Passive ICDs rely on either an orifice or a channel to create the pressure drop. For the autonomous ICDs, complex geometries are used to generate different flow behaviors, depending on the fluid properties. This allows the device to selectively restrict the flow of steam or water. The simplest design for an ICD is a nozzle or orifice (as shown in Figure 8.8). This will be discussed later in this chapter.

The flow behavior of ICDs is the result of the physics of the flow as defined by the governing equations. These equations are the starting point for understanding the flow. By solving a simplified form of the equations, the contract for differences (CFD) simulations are used to predict the flow in the devices. Then, by further simplifying the equations, a set of criteria is derived to quantify the performance of the devices. Different devices can then be compared using these criteria.

8.3.1 Fundamental ICD Design Equations

The design of ICD which has been discussed in earlier paragraph is related to software of a particular company but the basic physics of the flow through an ICD is governed by the conservation equations of mass, momentum, and energy. The fundamental differential form of these equations without simplifications is given in equations 8.1–8.4. For the continuity equation, the conservation of mass is expressed by:

$$\partial \rho / \partial t + \nabla \cdot (\rho \sim v) = 0, \tag{8.1}$$

where ρ is the density, t is the time, and ~v is the velocity vector. The conservation of momentum is given by:

$$\partial \rho \sim \mathrm{v} / \partial \mathrm{t}+\nabla \cdot(\rho \sim \mathrm{v} \sim \mathrm{v})=\nabla \cdot \mathrm{T} \sim+\rho \sim \mathrm{b}, \tag{8.2}$$

where ~b is the vector of body forces on the fluid and T~ is the stress tensor for the fluid given in index notation as:

$$\mathrm{Tij}=-\left(\mathrm{P}+2\ 3 \mu \partial \mathrm{uk} / \partial \mathrm{xk}\right) \delta \mathrm{ij}+\mu(\partial \mathrm{ui} / \partial \mathrm{xj}+\partial \mathrm{uj} / \partial \mathrm{xi}), \tag{8.3}$$

where P is the pressure, μ is the viscosity, and ui is the velocity vector in index notation. The conservation of energy equation is given by:

$$\partial \rho \mathrm{h}\ \partial \mathrm{t}+\nabla .(\rho \mathrm{vh})=\partial \mathrm{P}\ \partial \mathrm{t}+\nabla \cdot(\sim \mathrm{vP})+\nabla \cdot(\mathrm{k} \nabla \mathrm{T})+\mathrm{Q}^{\cdot}, \tag{8.4}$$

where h is the enthalpy of the fluid, k is the thermal conductivity, T is the fluid temperature, and Q˙ is the sum of the heat sources. Two types of fundamental design model of ICD will be explained in the subsequent sections.

8.3.1.1 Design, Installation, Procedure, Complexity, Cost, and Reliability

ICV technology is more complex; hence, ICDs have the advantage in terms of simpler design and installation, and lower costs. Although the simplicity of the ICD would imply greater reliability, there is little or no available operational data to support this, particularly when considering the greater likelihood of ICD plugging, due to scale, asphaltenes, waxes, etc., compared to ICVs. This initial study framework was extended to develop a comprehensive comparison of ICV and ICD application to an oil field in terms of reservoir, production, and cost engineering. The reasons behind the choices made are summarized as follows:

1. Uncertainty in reservoir description
2. More flexible development
3. Number of control zones
4. Tubing size
5. Value of information
6. Multilateral wells
7. Multiple reservoir management
8. Formation permeability
9. Long-term equipment reliability
10. Reservoir isolation barrier
11. Improved Clean-Up
12. Acidizing or scale treatment costs
13. Equipment cost
14. Installation: a) Complexity, b) Risk, c) Rig time

8.3.1.2 Reactive Control Based on "Unwanted" Fluid Flows of ICD

ICD completions restrict gas influx at the onset of gas breakthrough due to the (relatively) high volumetric flow rate of gas. Nozzle (orifice) type ICDs can also limit water influx due to the density difference between oil and water. However, an ICD's ability to react to unwanted fluids (i.e. gas and water) is limited compared to that of an ICV, especially a multi-set point ICV. ICVs allow the well to be produced at an optimum water or gas cut by applying the most appropriate (zonal) restrictions that maximize the total oil production with a minimum gas or water cut.

Proactive Control: ICD completions impose a proactive control of the fluid displacing oil. However, it is not possible to modify the applied restriction at a later date to achieve an optimum oil recovery, even if measurements were available that indicated an uneven advance of the flood front was occurring. ICVs, with their continuous flexibility to modify the inflow restriction, have the advantage.

8.3.1.3 More Flexible Development of ICV

An ICV's downhole flow path's diameter can be changed without intervention, whereas for an ICD, it is fixed once it has been installed. The ICV thus has more degrees of freedom than an ICD, allowing more flexible field development strategies to be employed.

8.3.2 Typical Type of ICD (Tesla Diode Model)

One of the devices is considered in ICD design based on Tesla's fluidic diode, shown in Figure 8.8. This device is designed to create a low pressure drop for flow in one direction, while producing a high pressure drop in the other direction. Originally designed by Nikola Tesla to act as a one-way flow valve without moving parts, the design would allow the valve to operate with high-temperature corrosive flows that would otherwise destroy a valve with moving parts. To increase the effectiveness of the device, Tesla proposed that multiple devices be placed in a series to maximize the one-way character of the device. Recent studies have adapted this device for use in mini-channels to provide a one-way flow valve. For this application, the Reynolds number at the entrance to the device is small, and laminar or transitionally turbulent flows are expected. Studies of Tesla's diode had considered on optimizing the diode properties of the device, as defined by the diodicity, the ratio between the pressure drop in opposite flow directions. This device has not previously been used in the design of ICDs. To consider this as an ICD, the device is used in the reverse direction so that the device creates a large pressure drop.

8.3.3 CFD Models for ICD Design

This type of simulation is performed using ANSYS-CFX 17.1 software. This software provides the advanced meshing capability needed for the complex designs that are considered. For the mesh, it uses an element-based finite volume-based method to discretize the governing equations given in equations 8.5 and 8.7. Meshing, as explained in the ANSYS theory guide, a finite element-based approach, is used to produce the mesh. In this method, the volume of the domain is divided into finite elements of various types. For the main body of the flow, the mesher defaults to tetrahedral elements, while for uniform flow areas, hexahedral element can be selected. In the inflation layer at the wall, prism and hexahedral elements are used to capture the boundary layer in the flow. Although the mesh is created with elements, the value of the solution field is stored at the nodes that form the corners of the element. By using shape functions for the elements, the value of the solution field at any point in an element can be calculated with a trilinear interpolation from the values at the nodes of the element.

8.3.4 Discretization Model of ICDs

For the discretization of the governing equations, the integral form of the equations is used on control volumes surrounding each node in the mesh. To estimate the values at the surface of the control volume, the element shape functions are used to linearly interpolate the value. To discretize the advection, it is necessary to estimate the value of variables at integration points around the control volume. The values at these integration points are then used to calculate the surface integral around each control volume. A variable, φip, can be estimated at an integration point using the following formula:

$$\varphi \text{ip} = \varphi \text{up} + \beta \nabla \varphi \sim \text{r}, \tag{8.5}$$

where φup is the variable value at the upwind node, β is a blending factor, $\nabla\varphi$ is the average gradient of the variable, and ~r is the vector from the upwind node to the integration point. For a blending factor of $\beta=0$, the scheme is a first-order accurate upwind differencing scheme (UDS). For a blending factor of $\beta=1$, the scheme is a second-order accurate central differencing scheme. While a UDS scheme is the most robust, it creates artificial diffusion and is only first-order accurate. This scheme is used for initial simulations to aid convergence, but a higher order scheme is needed for the final results. For the final simulations, the high-resolution scheme is used. This scheme uses an algorithm to maximize the value of β for each node to achieve a high-order solution, while avoiding artificial oscillations in the solution. Since this method uses a variable value for β, the order of the error is not consistent, and it cannot be used for mesh independence calculations. For the mesh independence simulations, a fixed blending factor of 0.7 is used to achieve a fixed-error order.

8.3.5 Type of Fluid Dynamics of ICDs

Once the flow leaves the reservoir and enters the production well, it passes through a variety of different flow regions before it is incorporated into the main flow inside the inner casing. In the initial sections, as it passes through the tubing and the annulus, the flow rate is low. As the flow enters the tubing, the flow rate increases resulting in turbulent flow. To understand the expected flow behavior in these regions, previous research into these flows is examined. For each region, the Reynolds number of the flow can be used to predict the expected type of flow. The Reynolds number is defined by:

$$\text{Re} = \text{Dvro}/\mu, \tag{8.6}$$

where Re is the Reynolds number, ρ is the fluid density, u is the average velocity of the flow, D is the diameter, and μ is the fluid viscosity. For non-circular geometry, the hydraulic diameter is used instead. The Reynolds number gives the ratio of the inertial and viscous forces in the flow. As the flow enters the outer liner, it passes through narrow slots in the liner that provide sand control. Since the overall production is spread out over all the slots, the flow rate through each individual slot is small. With the low flow rates, the Reynolds number of the flow in the slot is below one and creeping flow is expected. The hydraulic diameter of the slot is given by:

$$Dh = 4ab/2(a+b), \tag{8.7}$$

where a and b are the length and width of the slot, respectively. For a straight slot, a fully developed profile develops across the slot, when the depth of the slot is greater than the slot width. As the flow exits the slots, it enters an annular region between the slotted liner and the inner tubing. Since the slots, are evenly distributed in the outer liner, the flow rate in the annular region continuously increases as the flow gets closer to the ICD and more flow is added from the slots. As the flow enters the device, the flow rate rapidly increases. To understand the flow in the device, the models for simple devices are examined. The physics of the flow through basic orifices and channels has been discussed in detail, and pressure losses can be modeled using the Bernoulli equation with losses for internal flow, given in equation 8.8:

$$P1/\rho + V^1 2/2 + gz1 = P2/\rho + V^2 2/2 + gz2 + \left(f\,L\,D + \Sigma\,KL\right)\,V2/2, \tag{8.8}$$

where P is the pressure at two points, V is the velocity at two points and in the device, z is the height at two points, ρ is the density, g is the gravitational acceleration, f is the Darcy friction factor, L is the length of the flow path, D is the diameter, and PKL is the sum of the minor loss coefficients.

The plot of Fanning friction factor values changes with the change of Reynolds number and is shown in Figure 8.9, both in laminar and turbulent flow regime, which is not linear. The pressure losses are divided into minor losses, resulting from changes in flow geometry, and major losses from friction between the fluid and the wall. For the case of passive ICDs and single-phase incompressible fluids, the pressure loss behavior of the flow is often predicted using the minor losses and frictional losses from equation 8.8.

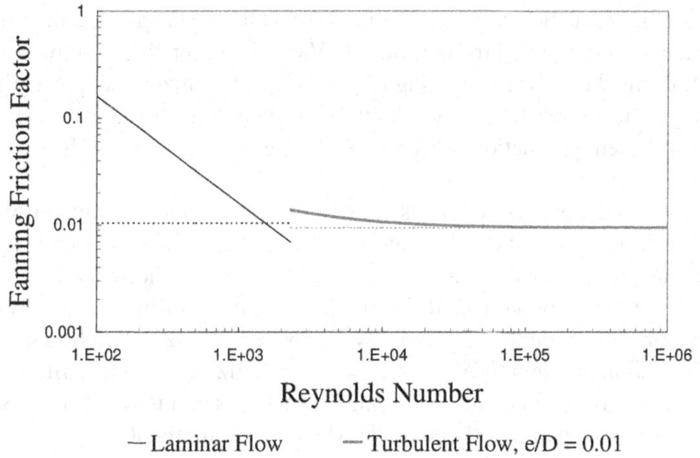

FIGURE 8.9 A plot of Fanning friction factor vs. Reynolds number.

By breaking down the device into a series of nozzles, bends, and channels, the minor loss coefficients and friction factor for each section can be estimated. Once the coefficients are known, the sum of the losses can be used to predict the behavior of the device. This drop in pressure depends on the strength of the vortex. For the pressure drop in the Tesla diode, two proposed mechanisms are outlined in Bardell's study. The first describes the pressure drop in terms of the momentum of the jet at the intersection where the channels rejoin. By performing a momentum balance for a control volume at this intersection, the pressure loss can be estimated. The second approach views the impingement of the jet from the loop as creating a vena contract in the flow at the channel intersection. This contraction of the flow then produces a pressure loss.

8.3.6 Laboratory Observation on CFD Models

Some lab results show that the unified model compares well with the CFD results. The average error percentage between the model and CFD results for flow distribution is 10.02%, while that for pressure drop is 11.25%. Both the model and CFD results show that the flow distributions in different paths of the splitter will be adjusted automatically according to the fluid's specific property, and thus, different fluids will enter the restrictor differently, and result in varying flow resistances. Specifically, oil, being more viscous, tends to take the restrictive path, enter the restrictor radially, and result in minimal flow restriction; while water, being less viscous, tends to take the frictional path, enter the restrictor tangentially, begin spinning rapidly near the exit, and result in obvious flow restriction. This autonomous function enables the well to continue producing oil for a longer time while limiting the water production; hence, the total oil production is maximized. The investigation conducted in this study also further enriches the theory of (https://www.sciencedirect.com/topics/engineering/hydrodynamics) hydrodynamic calculation for oil–water two-phase flow in complex.

8.4 INFLOW CONTROL VALVE DESIGN

Several production simulations model should be carried out to design and optimize the ICV sizing. Production rates vary from one field to another and are the function of the ICV size. In order to find the optimum ICV configuration that optimizes an objective function whether it is maximizing oil recovery, maximizing net present value, or minimizing water production, ICVs should be designed in such a way that each setting yields different production rate and thereby significantly impact the optimization process.

The ICV design process depends mostly on the production capability of each lateral. The design process depends on the average field production rate rather than individual lateral production rates. Pressure drop in the horizontal section also depends on the production rate and Fanning friction factor (which indirectly depends on production rate through the Reynolds number). ICV size was chosen to best fit each optimization problem. ICV settings were discretized, so each setting gave more or 20 less equal production rate even though sometimes that is impossible. Below are some considerations that have been made when designing the ICV:

- Simulation runs were performed to determine the maximum and minimum production rates for each lateral for two cases, i.e., when all laterals were producing together and when only one lateral was producing at a time.
- The area of an ICV was discretized based on the minimum and maximum production rate. The number of intermediate ICV settings is usually predefined by the manufacturing company. The minimum ICV setting corresponded to zero production rate, while the maximum ICV setting corresponded to the maximum production rate.
- If production rate was significantly different from one lateral to another, ICVs with different sizes were applied, although this might not be a feasible approach for all oil companies.

Selecting an ICD nozzle size with an optimal water breakthrough time and best recovery including an ICD in a horizontal well completion and reducing the nozzle size for fluid inflow automatically reduce fluid off take rate and delay the time when water enters the completion. The major focus is therefore to find an ICD nozzle size that delays water breakthrough, gives best off take rate, and gives the best ultimate recovery. Selecting an ICD nozzle size with minimal skin for effective ICD design and to ensure that the included ICD completion does not compromise production and effective recovery; a new equation that computes skin due to fluid entrance into ICD completions will be formulated to ease the ICD design process.

The modeling result helped in deriving equation of skin due to restricted fluid entry through ICD nozzles; the approach used is to assume radial flow into small nozzles of an ICD-equipped horizontal well in order to derive an approximate formula for calculating the skin or productivity loss due to restricted fluid movement into the horizontal well. Using the model in Figure 8.22, the fluid is bounded by the reservoir with boundary radius, re, having an average reservoir pressure, Pe. The fluid flows into the horizontal well as shown in Figure 8.23. For the establishment of this equation, Darcy radial flow into two regions is considered: (a) radial flow from the reservoir boundary into the flow impaired region (S) represented by the Darcy law:

$$\text{þe-þs} = qB\mu/2\pi kh \ln(re/rs), \tag{8.9}$$

And (b) radial flow from the impaired region (S) into the wellbore as represented by Darcy in (8.9):

$$\text{þs-þw} = qB\mu/2\pi k1h \ln(rs/rw). \tag{8.10}$$

The fluid flow through an ICD constriction/nozzle with radius "r" has a Darcy velocity represented by Darcy in the equation below:

$$V = q/\upsilon = Q/2\pi h \tag{8.11}$$

In equation 8.12, the nozzle flow area is the lateral surface area of a cylinder. "Since the interest is in determining the reduction in rate between flow through A and As in a horizontal well, h is assumed to be radius, r." Equation 8.5 is substituted into equations 8.3 and 8.4 and transformed into equations 8.6 and 8.7:

$$ \text{þs-þw} = qB\mu/2\pi k1rh \ln(rs/rw). \quad (8.12) $$

Substitute the flow area of the nozzle A = into equations 8.14 and 8.15, the equations become, Adding

$$ \text{þe-þs} = qB\mu/2\pi krh \ln(re/rs) \quad (8.13) $$

Adding equations 8.14 and 8.15 transforms into:

$$ \text{þs-þw} = qB\mu/2kAs \ln(rs/rw) \quad (8.14) $$

$$ \text{þs-þi} = qB\mu/2kA \ln(ri/rs). \quad (8.15) $$

According to Darcy model:

$$ \text{þe-þw} = qB\mu/2\pi kh \ln\sum \ln(ri/rw)/A1 + \ln(ri/rs)/A). \quad (8.16) $$

According to Darcy skin model:

$$ \text{þe-þw} = qB\mu/2\pi kh(\ln(ri/rw)+S). \quad (8.17) $$

According to skin model, in a horizontal well with ICD, skin model can be regenerated into:

$$ \text{þe-þw} = qB\mu/2k\,A(\ln(re/rw)+S). \quad (8.18) $$

Elimination of total pressure drop, Pe-Pw, between equations 8.10 and 8.12 and further simplification of equation 8.20 degenerate into:

$$ S = (A/A1\text{-}1)\ln(ri/rs), \quad (8.19) $$

where S is the the pseudo-skin due to restricted fluid entry into ICD nozzles, A is the equivalent nozzle area of an ICD less horizontal well, which is 0.002 ft^2 from analysis conducted, as is the average nozzle area for all ICD joints in a horizontal well, and rw is wellbore radius; rs = (1−average nozzle radius of all ICD joints)/2.

8.4.1 Ideal Characteristics of Performance Criteria of ICDs

When designing new ICDs or comparing current designs, the criteria should be linked to the performance of the ICDs and be independent of the flow conditions. To determine the quality of criteria, a standard is developed that details the ideal characteristics of a set of criteria. By comparing criteria with this standard, the quality of the criteria can be evaluated. An ideal set of criteria should be:

- Independent of the flow rate
- Independent of the fluid properties
- Independent of the scale of the device
- Quantifiable value that can be easily measured
- Linked to specific aspects of the performance
- Adapted for use in a design optimization. Independence from the flow rate is important, since the flow rate through the device varies greatly over the life of the well, depending on the fluid being produced.

In connection with this, the criteria should also be independent of the properties of the fluid being produced. Ideally, the criteria are exclusively dependent on the design of the device. In this way, the overall performance of the device can be quantified. Even for a single ICD design, the device is scaled by adjusting the dimensions to achieve a particular flow resistance. To prevent this device scaling from affecting the values, the criteria should be independent of the scale of the device. Although the criteria should not depend on all dimensions, there are dimensions inherent in the design that should be included in the criteria. In calculating the criteria, precisely quantifiable values should be used that allow for their detailed comparison. Since there is a wide range of designs, the criteria need to be defined, so they are not restricted to only a few types of designs. For this, the criteria should be based on the performance for a wide range of flow conditions. At the same time, the criteria should be calculated from a small set of data, to avoid extensive simulations or testing. In reality, criteria that meet precisely all these conditions cannot be found. But by comparing criteria to this standard, the quality of the criteria can be evaluated. Based on these ideal characteristics, current criteria can be evaluated and an improved set of criteria can be developed. Although multiple criteria allow for an informed comparison, the lack of a single value makes it challenging to select an optimum design. To resolve this issue, the criteria should be designed for an optimization. Then, the criteria can easily be used in a design optimization like that proposed by Li et al. for outflow control devices. In this type of optimization, various design parameters are varied, using design of experiments, to find the design that minimizes the criteria.

8.4.2 Limitations of Current Criteria of Selection

In order to evaluate the current criteria, they are compared with the ideal criteria characteristics. Current criteria for ICDs are used to provide a quantitative measure of the flow resistance of the device. The most basic measure is the flow resistance rating, defined as the pressure drop in bars for water at a flow rate. While this is useful for sizing a device, it is dependent on the scale of the device and cannot be used to compare designs. As a single pressure drop value, it does not give any information on the behavior of the device. The most commonly used criteria for analyzing the behavior of ICDs are the discharge coefficient, Cd, defined as:

$$\Delta P = \rho u2\ 2C2\,d, \tag{8.20}$$

and the total loss coefficient, Kt, defined as:

$$\Delta P = Kt\ \rho u2\ 2. \tag{8.21}$$

These are relatively independent of flow rate and fluid properties for orifices, when the Reynolds number is large. But, when the Reynolds number is small or the device has significant frictional losses, they become dependent on the flow rate and fluid properties. This dependence means that the criteria can only be compared at similar flow conditions. Because of this dependence, they do not fully describe the behavior of the devices. The reason for this dependence is that the pressure losses are the combination of two types of losses. In the previous paragraph, the pressure loss behavior of ICDs was expressed using the Bernoulli equation, where the losses are given by:

$$\Delta P = 1\ 2\ K + fL\ D\ \rho u2. \tag{8.22}$$

For these criteria, the minor and frictional losses are combined into a single coefficient, as either the total loss coefficient or the discharge coefficient. By including the frictional losses in the coefficient, the coefficient becomes dependent on the flow rate and fluid properties, since the friction factor depends on the Reynolds number of the flow. Due to this dependence, the coefficients are often expressed as a function of the Reynolds number in the device. While it would be convenient to express the losses as a function of only the Reynolds number, it is not this simple. Although the friction factor is a simple function of the Reynolds number (Figure 8.11), the rest of the loss equation shows a separate dependence on the density and flow rate, resulting in a complex dependence. Since the discharge coefficient and total loss coefficient are not fixed, a comparison of the coefficients has a limited scope. To allow for a broader comparison, plots of the flow behavior are often used to qualitatively compare devices. To account for the various aspects of the device performance, multiple criteria are needed. By making the criteria more specific, the dependence of the criteria on the flow conditions is reduced. In this way, minor and frictional losses can be evaluated with separate criteria, developed to minimize the flow rate and viscosity dependence.

Flow control technology is the "hands" of intelligent well technology. The ultimate goal of intelligent well technology is to regulate and reorganize oil recovery % with respect to probability of density variation with respect to pressure, which is shown in the graph in Figure 8.10.

Adaptive inflow control device (AICD) is the conventional ICD that does not have the late period adjustment capacity, but the ICV is expensive, and its long-term reliability and control range are limited, which are only suitable for oil and gas wells with high output and long production cycle. Therefore, it is desirable to develop a kind of water control device that can evenly control the liquid in the early stage and realize water blocking-oil recovery in the late period. In order to solve this problem, major oil companies at home and abroad have provided different types of AICD. Technically, by means of materials, spring, sliding block, etc., such devices are able to achieve the high resistance to water/gas and low resistance to oil. The typical AICD devices are rate-controlled production (RCP) valve developed by Statoil, which relies on the "floating disk" to block water and stabilize oil, and the AICD developed by

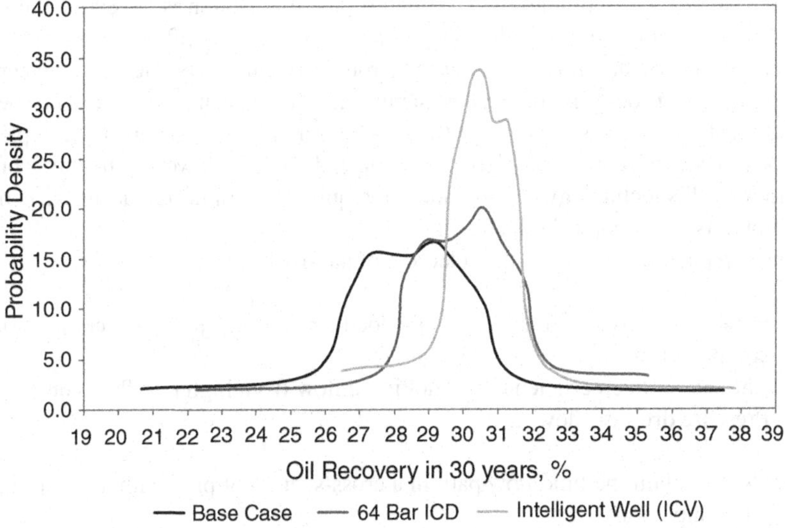

FIGURE 8.10 Probability density with respect to various pressure values vs. recovery of oil over the years.

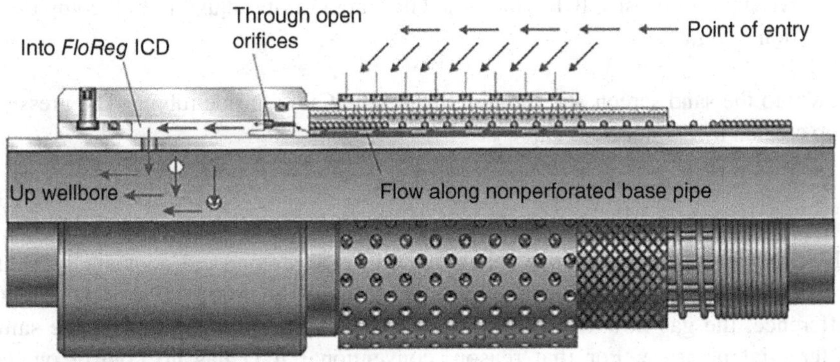

FIGURE 8.11 Schematics of orifice type ICD.

Halliburton, which relies entirely on "flow-path conversion" to block water and stabilize oil. Adaptive, low cost, high intelligence, and high reliability will be the development trend of the AICD in the future (Figure 8.9). In central tube completion water control technology, central tube completion is to insert a tubing string (in most cases, with a packer) smaller than the wellbore diameter into the conventional completion wellbore. After the central pipe is inserted, the horizontal section can be divided into several parts according to the liquid production, and the fluid flowing inside the wellbore also changes into three parts, namely, flowing in the annulus, flowing in the wellbore, and flowing inside the central tube. During conventional completion, due to reservoir heterogeneity and wellbore pressure drop along the horizontal section, the inflow profile is severely uneven. The central pipe completion is to reduce the pressure drop of the high-permeability section and re-match the inflow pressure drop in the central tube, making the entire inflow profile more uniform. Because the central pipe completion process is simple, cost-saving, and convenient, and there is no special requirement on the tools, it is one of the most commonly used secondary completion methods for water control completion at home and abroad. However, the fluid control accuracy of this technology is poor, and the requirement on design accuracy is high. Therefore, its water control effect is limited.

The three main criteria for the selection of passive ICD are as follows:

- Flow restrictive element is embedded into each joint of completion equipment.
- Enhances sweep efficiency by choking inflow from high rate flow zone.
- Pressure drop is ~ flow rate.

Figure 8.11 explain the fluid flow path in a cross-section of pipe with a nozzle-based ICD well completion,

- Pressure drop created by the nozzle is explained by the empirical formula given below.
- All types of passive ICD. Have fixed flow area limits adjusting well completion design

Flow into the sand screen and through the nozzle ICD to inside tubing. The pressure drop created by the nozzle is expressed by:

$$\Delta P = C \cdot \rho \cdot v\ 4, \tag{8.23}$$

where ΔP is the pressure drop across ICD, C is a geometrical constant, ρ is the produced fluid density, and v is fluid velocity through the nozzle. Due to the density difference, the gas flow rate is significantly higher than the oil flow at the same differential pressure. For that reason, conventional ICD has no control on gas production.

The performance curves of the nozzle ICD for oil, water, and gas. All passive ICDs have fixed-flow area which limits adapting completion design in case of unexpected reservoir fluids saturation, pressure and permeability profiles encountered. Also, after putting well on production and gas/water breakthrough, passive ICDs did not give a hand to control gas production as shown in Figure 8.12 and only control water production to a certain degree and not complete water shutoff. There is a need of complete undesired fluid shut off push to develop On/Off ICD type which is equipped with sliding sleeve. This type of On/Off ICDs enables operators to complete shutoff compartments that produce undesired fluids. Many factors limited the spread of On/Off ICDs application such as production logging tool (PLT) requirements to identify where gas/water breakthrough, using Coil Tubing/ Tractors with shifting tool to shut off ICDs.

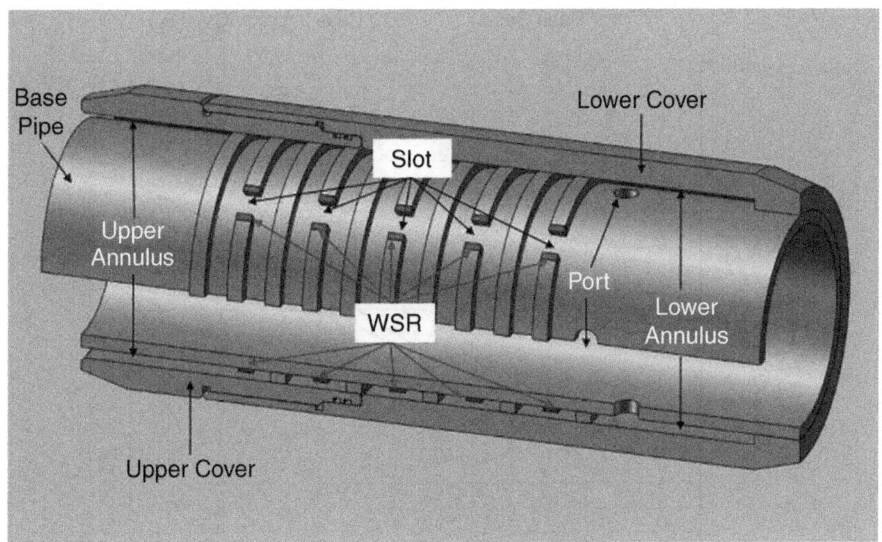

FIGURE 8.12 Autonomous inflow control valve.

8.4.3 Autonomous Inflow Control Valves

Autonomous inflow control device (AICD, Figure 8.12) design is proposed based on the combination of two fluid dynamic components, with the splitter directing the flow and the restrictor restricting the flow. Based on the combination of the flow pattern transformation criterion, homogenous model, two-fluid model, and pipe serial–parallel theory, a unified model of oil–water two-phase flow is developed to predict both the flow distributions and pressure drops through the splitter, which is then compared with the computational fluid dynamics (CFD) results. Also, the rules of oil–water two-phase flow through the disk-shaped restrictor are studied by numerical simulation.

8.4.3.1 The Need of Autonomous ICV

The novel AICV technology can differentiate between fluid types based on viscosity and density, and when undesired fluid (gas and/or water) production starts, the AICV starts to choke the valve flow area gradually until completely shut off without well intervention. Inflow control device AICD becomes essential, where ICD can detect the type of produced fluid and have movable parts enable choking/shut-off ICD flow area when undesired fluids are produced. There have been many attempts to develop AICD; the AICD that has movable parts working according to the type of fluid produced was first installed in 2007. The AICD developed by Statoil is the RCP valve in troll field.

Fluid flow performance curve of an RCP compared to nozzle-based ICD is shown in Figure 8.13, where the flow rate of oil is equal for both devices. Due to the change in flow area for different fluids, the flow rate of gas and water will

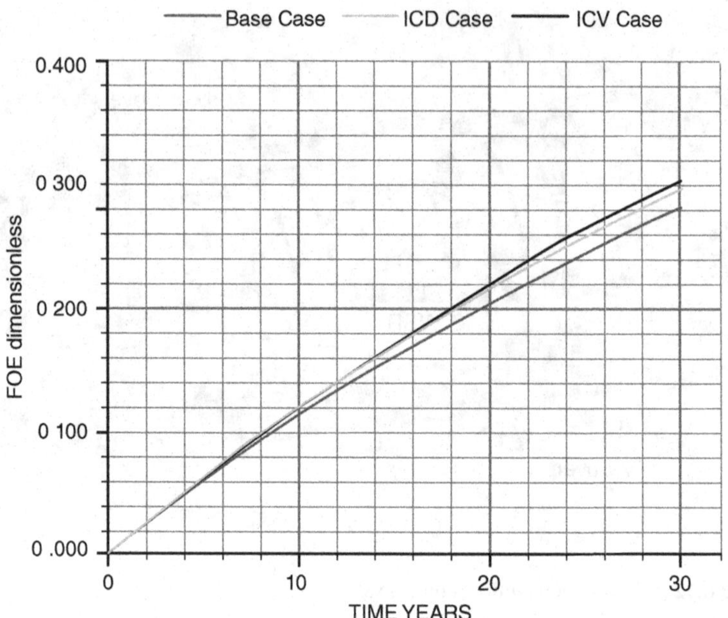

FIGURE 8.13 Fluid flow performance curve.

be less for the RCP compared to the passive ICD type. The black curved lines in the figure show the RCP improvement in the flow performance for gas and water compared to passive ICD. Due to lower flow rate for low viscous fluid, the gas flow rate will be lower than that for the passive ICD. The improved flow performance curve for gas will reduce the well GOR. For the autonomous ICD designs, other types of flow behavior create additional pressure losses in the devices. In vortex-based designs, like similar Halliburton's fluidic diode, the fluid causes a vortex as it tangentially enters the circular region. This vortex takes the form of a compound or Rankine vortex, where the outer part of the vortex is irrational and the core of the vortex rotates as a solid body as it exits the region. This results in an increase in the tangential velocity of the flow, as it approaches the core with a sudden decrease at the core. As the velocity increases, the pressure decreases by Bernoulli's equation (Figure 8.8).

8.4.3.2 ICD Flow Selection

Passive ICDs were introduced to balance fluid influx across open-hole section and control water production. ICDs have successfully demonstrated that they can delay the gas and/or water breakthrough in horizontal wells, but they cannot choke the gas production effectively after coning/breakthrough and are limited for water production control. Figure 8.11 shows a pipe cross-section with nozzle-based ICD completion, where the red arrows illustrate the fluids path. The fluid flows to the well, from the annulus via the sand screen and into the nozzle ICD. When the fluid flows through the small nozzle, pressure drop is generated as a function of fluid velocity squared, density, and geometry of the ICD. The pressure drop is almost independent of fluid viscosity. The nozzle size and the pressure drop for a specific fluid developed by the device are set prior to installation. But once number/size of nozzle had been installed, it will remain as constant flow area even if type of produced fluid changes—passive system (Figure 8.12).

8.4.3.3 Flow Adjustment and Water Flow Control Completion Technology

Horizontal well flow adjustment water control completion technology is to install the downhole ICD to control the inflow of edge and bottom water effectively. According to the fluid inflow control modes, it can be divided into ICD, ICV, and AICD. ICD is a throttling device that creates additional pressure differential when the reservoir fluids flow into the wellbore through this device. By adjusting the type of throttling device to increase the additional production pressure drop of the high-permeability section (reducing the effective production pressure drop) and to reduce the production pressure drop of the low-permeability section (increasing the effective production pressure drop), it is able to eliminate the gas/water coning caused by reservoir heterogeneity and the heel–toe effect, so that the horizontal well production liquid profile is balanced, and the oil–water interface is uniformly raised. Major oil companies at home and abroad have independently developed different types of ICDs. According to different throttling methods, they can be divided into three categories: spiral ICD, nozzle ICD, and flow-path ICD. In addition, the special ICD with spring and slip block (Figure 8.9) has also been discussed, but the application is limited. Because the ICD realizes the throttling through adjusting the passage of fluids, it is generally connected with the screen, which is usually called the flow adjustment water control screen. Such screen has been widely used at home and abroad. In recent years, the flow adjustment water control screen completion technology has been widely applied for its low cost, simple operation, and good water control effect. AICV completions can be considered an optimum completion solution to overcome the Fahud Assets reservoir and production challenges. AICV advances passive ICD technology to another level by making AICV valve smarter and hosting features of dynamic flow choking of unwanted phases (which is more comparable to intelligent completions flow control valves), yet the AICV does this without any control lines to surface. The AICV is a truly autonomous valve, as compared to the passive ICD (which cannot change its effective flow area to stop gas and water); the AICV technology has a robust, dynamic piston that can close and reduce the unwanted gas and water production.

8.4.3.4 Inflow Control Valve

Since reservoir properties and phase (oil, gas, and water) saturation change over time, it is desirable to have a downhole flow control technique with post-regulation capability that can be automatically adjusted based on actual inflows. The intelligent well technology that emerged in the 1990s provided a means to solve this problem. One of the key components in intelligent well technology is the ICV. The working principle of the ICV is equivalent to one downhole nozzle, and the flow rate of each section is adjusted by downhole throttling, and thereby, the purposes of adjusting the inflow profile and delaying water breakthrough can be realized. In addition to downhole installed ICV, a complete ICV system shall be equipped with pressure/temperature (PT) sensors to monitor and analyze the downhole production status in real time and then transmit the resulted control information to downhole control unit to adjust the ICV through the control mechanism. Several major foreign oil companies can offer different types of ICV. The most advanced electronically controlled "infinite" adjustable ICV combined with position sensor can provide valve position control of over 100 opening degrees; the simplest ICV owns the limited discrete valve positions or only two positions of opening and closing. ICV is typically powered by the hydraulic, electric, or electro-hydraulic system. At present, the hydraulic control technology is relatively mature, and one hydraulic system can control up to 8 ICVs. Compared with ICD, ICV is an "active control" device. After well completion, according to changes in downhole production status, the opening degree of ICV can be adjusted timely through the surface control system to achieve the purpose of real-time production optimization. Compared with other downhole ICDs, ICV is more suitable for dealing with the impact of uncertainties from reservoir properties in the production process. It owns more flexible development method and has good adaptability in low-permeability formations. It has strong acidizing/fouling treatment capacity and good water control effect, and it is able to realize the optimal production of heterogeneous reservoirs or multilateral well. However, ICV is inferior to ICD in terms of long-term reliability and control scope.

8.4.3.5 Interval Control Valves Case Study-1

In this typical case study five no-producing zones have been considered for ICV selection, which is explained as follows:

- 5 ICVs divide 2,527 m completion interval into zones
- 1st, 3rd, 4th, and 5th zones in layer
- 2nd zone in layer
- Valves assumed to be infinite variable ones
- 3.5″ tubing (required for ICVs in 8.5″ hole)
- Tubing Head Pressure allowed 600 m^3/d production at all times
- ICVs operated to limit production from zones with high WOR
- Gas cap is weak, and gas choking does not increase recovery.

Passive ICDs were introduced to the market in mid-90s, to balance fluid influx across open-hole section and control water production. ICDs have successfully demonstrated that they can delay the gas and/or water breakthrough in horizontal wells, but this cannot choke the gas production effectively after coning/breakthrough, and they are limited for water production control.

It can be explained from Figure 8.14 that with respect to temperature, the fluid flow behavior of three different flow regimes, i.e., base case, ICD, and ICV, changes with time (Figure 8.15).

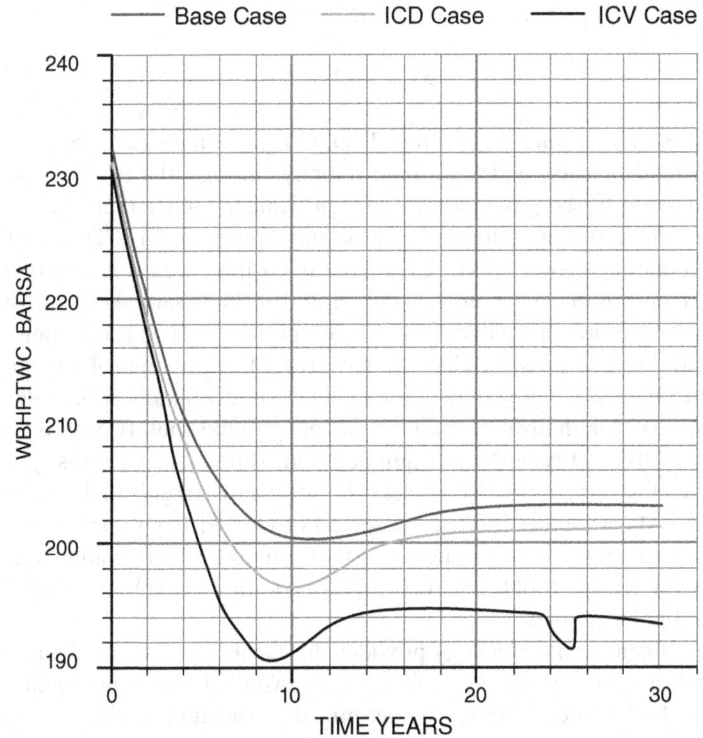

FIGURE 8.14 Fluid flow behavior in three different flow regimes.

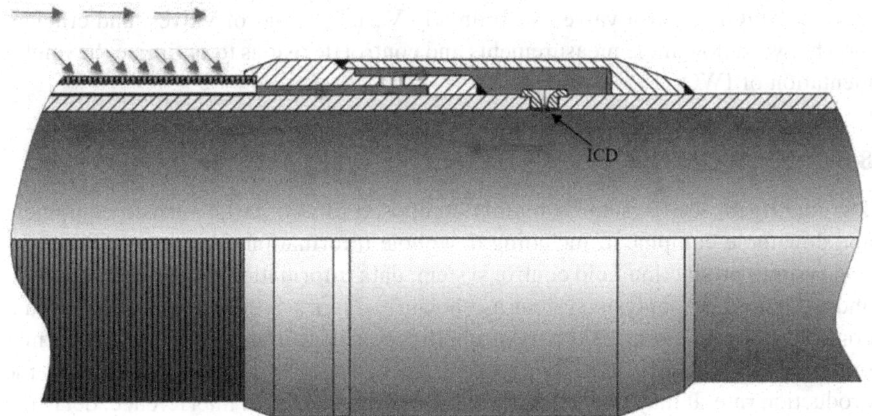

FIGURE 8.15 Flow into the sand screen and through the nozzle ICD to inside tubing.

The pressure drop created by the nozzle is expressed by:

$$\Delta P = C1/2\,\rho v2, \tag{8.24}$$

where ΔP is the pressure drop across ICD, C is a geometrical constant, ρ is the produced fluid density, and v is fluid velocity through the nozzle. Due to the density difference, the gas flow rate is significantly higher than the oil flow at the same differential pressure. For that reason, conventional ICD has no control on gas production. All passive ICDs have fixed-flow area that limits adapting completion design in case of unexpected reservoir fluids saturation, pressure, and permeability profiles encountered. Also, after putting well on production and gas/water breakthrough, passive ICDs did not give a hand to control gas production as shown in Figure 8.14 and only control water production to a certain degree and not complete water shutoff. A complete undesired fluid shutoff push is needed to develop On/Off ICD type that is equipped with sliding sleeve. This type of On/Off ICDs enables operators to complete shutoff compartments that produce undesired fluids. Many factors limited the spread of On/Off ICDs application such as PLT requirements to identify where gas/water breakthrough, using Coil Tubing/Tractors with shifting tool to shut off ICDs, in addition to the high Cost/Risk from operational point of view.

The intelligent well technology provides the capability to remotely monitor and control multiple production zones using ICVs installed on the production tubing. Here, a method for determining the optimized application of different intelligent well control strategies is presented. The optimization algorithm, which is based on trust region method, has been coupled with a commercial flow simulator and applied to a conceptual sandstone thin oil rim model containing a heterogeneous high permeable longitudinal channel to quantify the benefits of intelligent completions over a base case with a conventional completion. Three control strategies have been defined, analyzed, and optimized, including a Fixed-Flow Control Device, On/Off control valve and Infinitely Variable Control Valves, and efficient loop between downhole measurements and control decisions to optimize the implementation of IWCs.

8.5 COMPOSITION OF INTELLIGENT WELL TECHNOLOGY

The intelligent well system is mainly composed of two parts: surface equipment and downhole equipment, including downhole information detection and acquisition system, production fluid control system, data information transmission system, and up-hole data analysis system as shown in Figure 8.16. The production fluid control system is an integral part of intelligent well technology. When the formation pressure is insufficient, the reservoir energy can be restored by adjusting the production rate. It may thus effectively control the interlayer interference, delay the water breakthrough, inhibit the water content, prolong the high-efficiency exploitation time of the reservoir, optimize production of the oil well, and enhance oil and gas production.

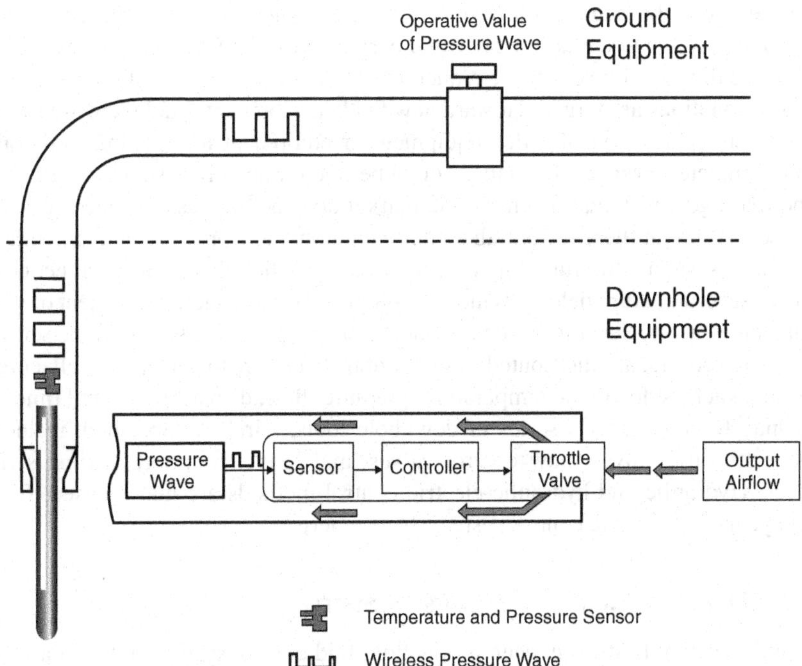

FIGURE 8.16 Logic flowchart of downhole data monitoring system of an intelligent well.

8.5.1 Downhole Information Detection and Acquisition System

The digital oil field concept has been used in temporary applications such as drilling and well test operations with great success. However, the uptake in permanent completion applications has been little low, and even those that have been implemented are often not utilized to the maximum effect.

Significant advancements have recently been made on surface systems, partially due to the overlap with other industries and the role Big Data plays in the digital oil field vision. Downhole components have unfortunately not progressed at the same speed, and new advancements in downhole monitoring and control technology are required to provide in-well production optimization.

The early digital oil field allowed companies to capture more data and have those analyzed in real time or near real time to optimize production and improve well performance. One of the key attributes to this was improving efficiency in decision-making by removing the middle man and delivering the data directly to the right people and into the right models. The vision now is that the digital oil field will take the next step and provide full automation for production optimization by using truly intelligent systems and closing the feedback loop.

Most intelligent downhole monitoring and the control equipment used are operated via hydraulic and/or electric control lines that run between the downhole device and surface control system to provide power and communication. These control lines can often limit the completion design and lead to increased cost and complexity. They also have the potential for early and permanent failure, meaning a key component of the digital oil field may be reductant before the well is brought online.

These solutions are targeted toward new field developments, and there are limited options for replacement of failed equipment or equipment for existing wells other than a complete workover. The same cannot be said for downhole solutions, and most of the intelligent well equipment on the market does not address the needs of existing assets. Without these retrofittable intelligent downhole pressure and temperature transmitter systems, the full benefit of the digital oil field is out of reach economically for several mature fields. It's mainly composed of downhole permanent or semi-permanent sensors. Several sensors, including electronic sensors, fiber-optic sensors, and quartz sensors, are distributed throughout the wellbore to detect and collect real-time data such as downhole temperature, pressure, flow, displacement, and time.

It mainly consists of a series of downhole tools, wireless controlled, including packers, throttle valves, flow control valves, and branch wellbore sealing switch devices. Hydraulic and hydraulic-electric control methods are mainly used. But in recent years, all-electric control systems have emerged.

8.5.2 Data Information Transmission System

It is an important bridge between production fluid control system and data analysis system. It plays the role of communication between real-time downhole data acquisition and transmission of control instructions to downhole. Generally, it is realized by cable or optical fiber. In recent years, it also transmits information by means of pressure pulse or electromagnetic wave decoding. The measurement of intelligent well technology is continuous and steady. Figure 8.16 shows the main logic diagram that controls the reservoir monitoring software to generate real-time bottom-hole raw data such as pressure and temperature through bottom sensors as explained in Figure 8.16.

Compared with traditional well completion technology, intelligent well technology has significant advantages of accuracy, continuity, convenience, and efficiency.

8.5.2.1 Real-Time Downhole Condition Monitoring and Data Acquisition

This system provides continuous steady production logging data.

It helps to ensure higher accuracy, avoid the impact of traditional logging on normal production, and secure higher efficient including surface data.

8.5.3 Up-Hole Data Analysis System

It's used to process the raw data collected by the downhole sensor and is summarized and screened by the software system to help the user to clearly understand the downhole occurrence. At the same time, the analysis software adopts reservoir engineering method, optimization method, and reservoir numerical simulation and prediction technology to analyze and collect the production dynamic data and help users make corresponding adjustment decisions.

8.5.4 Characteristics of Intelligent Well Technology

For the implementation of intelligent well technology, continuous monitoring of reservoir and production engineering management is required. Through continuous, long-term, real-time, and accurate downhole information data, the ground system simulates and analyzes reservoir occurrence. It helps engineers to judge downhole conditions and make decisions, facilitate accurate management of production, improve production efficiency, and mitigate safety hazards.

8.5.5 Surface System Analysis and Decision-Making and Closed-Loop Control of Downhole Tools

Logging and control in traditional technology are two independent processes. They cannot be carried out at the same time. The well still changes continuously between the completion of logging and the start of production regulation, leading to the inaccuracy of traditional technology in production optimization. The test and control of intelligent well technology can be carried out simultaneously. It not only has advantages in decision accuracy but also reduces the workload and working time of logging and lifting tools, greatly improving production efficiency.

8.5.6 Surface Readout (SRO) Data from Bottom-Hole Pressure Transmitter

Measurements can be transmitted to the surface, usually via an electric cable/fiber-optic cable, or recorded in downhole memory powered by batteries. SRO has the obvious advantage of providing data in real time. Real-time readouts are especially beneficial for transient measurements that require time for the pressure to stabilize and radial flow to develop. Because stabilization times depend on reservoir and fluid properties and because the determination of these parameters is often the purpose of pressure measurements in well testing, predicting the duration of stabilization periods is often difficult. SRO is preferred in these cases. The drawback is the difficulty of guaranteeing the quality of the acquired data, including the potential for significant operating losses if the bottom-hole recording equipment malfunctions. For these reasons, downhole recording (DHR) should be chosen only when the measurement target does not necessarily depend on stabilization times or when stabilization times are already known (e.g., to measure the average reservoir pressure in a reservoir of known mobility).

8.5.7 Surface Readout Pressure Transmitter and Downhole Recorder

Many oil companies installed both SRO and DHR. The measuring section of these tools is common to both options. In the SRO option, the sensor electronics are coupled to a telemetry system for up-hole transmission, and the cable supplies downhole power. In the DHR option, downhole batteries supply power, and the data are stored in memory boards for future readout or downloading to suitable computer systems. The downhole data transmission system and electronic sensors

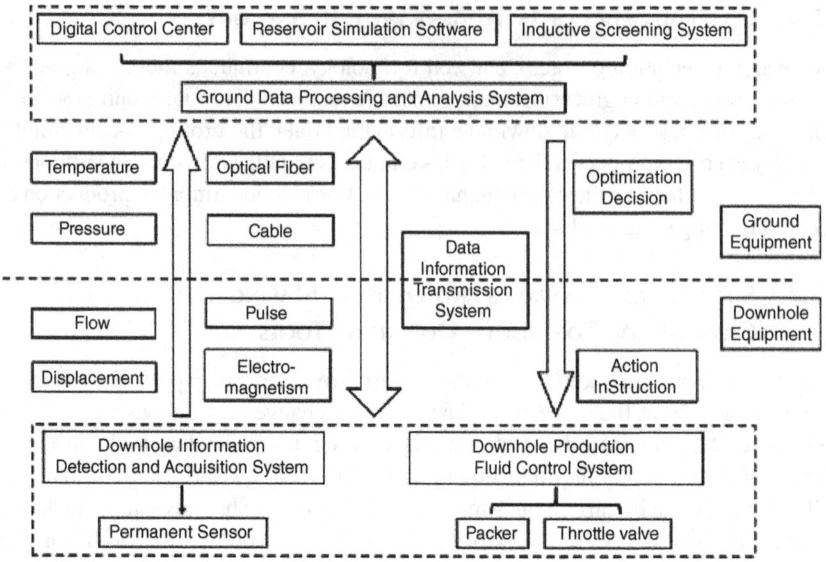

FIGURE 8.17 Continuous and steady measurement of intelligent well technology.

are independently developed by WRS, Promore, which can monitor downhole temperature and pressure and transmit them to the ground in real time. Continuous and steady flow diagram is shown in Figure 8.17.

The distributed FBG temperature and pressure sensors are divided into two kinds: single-point sensor and double-point sensor. Both of them can perform downhole temperature and pressure measurement. Single-point sensor can measure the temperature and pressure of tubing or annulus separately, and double-point sensor can monitor the temperature and pressure of tubing and annulus simultaneously. The supporting brackets of sensors are also divided into single-point type and double-point type, helping to better fix the supporting sensors and fully protect the sensors in the working process and effectively securing the safety and reliability of sensor signal transmission.

8.5.8 Downhole Swell Packer

This down hole equipment is used for the isolation of multiple zones for the completion of multitoned intelligent wells. The layer wise isolation is achieved by the activation of rubberized elements on the outside of swell packer by swelling. The swelling is achieved either by oil or water (Figure 8.18). Elastomers that swell in oil could result in a reduction of the designed performance. Over the course of time, new technologies, such as horizontal drilling, multistage fracturing, and reservoir management with the use of ICDs, required something different for annular isolation than cement.

With the introduction of non-cemented completions came the necessity of an open-hole packer. This opened up the opportunity for swellable packers. Natural rubber swells when exposed to hydrocarbon-based fluid. The amount of swell is dependent on the chemistry of the oil and the temperature at which the exposure occurs. Oil is absorbed

FIGURE 8.18 Diagram of swell packer.

into the oil-swellable elastomer through diffusion. Through the random thermal motion of the atoms that are in the liquid hydrocarbons, oil diffuses into the elastomer. When the elastomer is wrapped on a piece of pipe, the result is an increase of the manufactured outside diameter of the oil-swellable elastomer. Oil continues to diffuse into the elastomer causing the packing element to swell until it reaches the inside diameter of the open hole. The swelling continues until the internal stresses inside the elastomer reach equilibrium. That is, the swell pressure increases until diffusion can no longer occur. At this point, a differentially sealing annular barrier is created.

There are no moving parts in swell packer, through pipe manipulation or by applied hydraulic pressure. The packers are simply lowered to depth, similar to the casing, and allowed to swell before production or injection operations begin.

The elastomer is wrapped on American Petroleum Institute (API) pipe with weight, grade, and connection specified by the well design. The seal length is determined by the required differential pressure, the well temperature, and application. Solid metal end rings are secured to the base pipe and protect the elastomer, while the completion is run to depth. Critical to the function of the end rings is that they create an extrusion barrier. As differential pressure is applied across the packer, forces are created that want to extrude the rubber in the direction of the applied pressure. The end rings support the elastomer, which results in a hydraulic annular seal. Swellable packers can have pressure ratings as high as 10,000 psi.

For low-pressure applications, there is a slip-on version of the swellable packer. This packer works similar to a slip-on centralizer. It is often used to provide annular isolation for slotted liners, or flow barriers with sand screens that use ICDs.

Elastomers have also been developed that will swell in the presence of water and water-based fluids. Swell is achieved by blending in a super-absorbent polymer into the base elastomer compound. Once the packer is exposed to water, the water is absorbed by the polymer causing the packer volume to increase. This increase in volume corresponds to an increase in the packer diameter. Just like the oil-swellable elastomer, a seal is created once contact with the borehole wall is made. The water-swellable packer is capable of swelling in both monovalent and divalent cation brines. Using the super-absorbent polymer creates a very stable water swell packer. If the water is removed and replaced with crude oil, the packer will maintain a seal.

Recently, a water-swellable elastomer has been designed and installed in heavy oil wells, which use steam to heat up the reservoir to cause oil to flow. The high-temperature water-swellable packer presently has an upper operational limit of 575°F (302°C).

8.5.9 Bottom-Hole Pressure and Temperature Transmitter

The temperature transmitter consistently monitors the accurate pressure measurements with the Signature CQG crystal quartz gauge (Figure 8.19). By measuring pressure and temperature at a single point with a single quartz crystal, the Signature CQG gauge significantly reduces thermal effects. Small residual effects are further minimized with real-time dynamic compensation. This unique capability also speeds pressure response to truly capture dynamic downhole conditions.

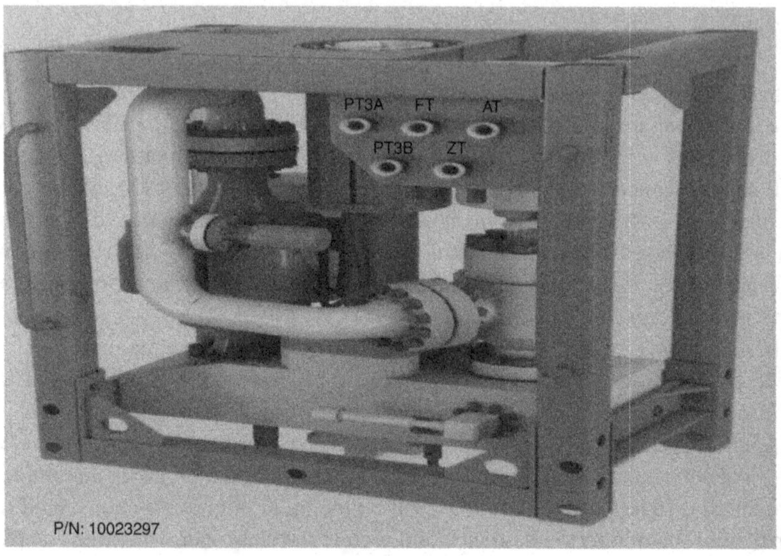

FIGURE 8.19 Bottom-hole wet gas meter (subsea well).

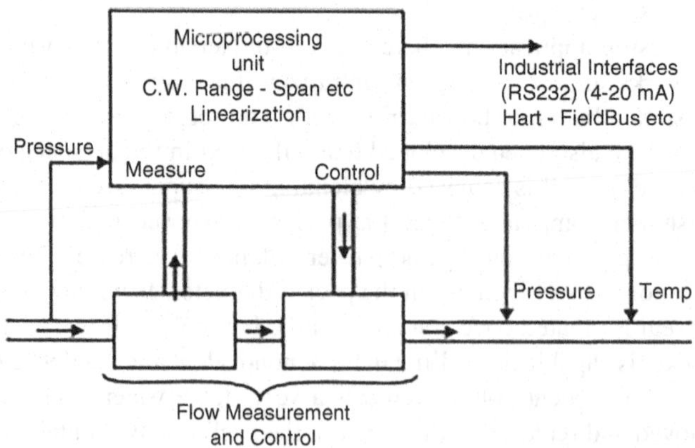

FIGURE 8.20 A microprocessor-based fluid flow loop sketch and measurement equipment.

8.5.10 Bottom-Hole Sonic Gauge

Wireless Monitoring System is a next-generation acoustic telemetry system that measures wellbore pressure and temperature in real time, a mandatory requirement for good reservoir management in oil and gas wells. Wireless downhole gauge systems are ideal for providing real-time downhole data during drill stem testing and production well testing. The technology can also be deployed in free flowing wells, wells equipped with artificial lift systems, and in observation wells. There is no depth limitation for the Sonic Gauge System as acoustic repeater stations can be used to boost the acoustic data packets to surface. Both the wireless sensors and repeaters can be run on tubing or retrofitted through tubing on high-expansion gauge hanger via traditional slick line or e-line. Sonar-based flow measurement technology was first introduced to the oil and gas industry with the deployment of the first downhole, fiber-optic-strain, multiphase flow meter from the Shell Mars Platform in October 2000. The fiber-optic-strain sonar meter used sonar-based passive listening techniques to provide flow rate and compositional information for downhole oil, water, and gas mixtures and served as a technology demonstrator for a product line of downhole, fiber-optic flow meters commercially available today. In 2004, a clamp-on version of a strain-based sonar meter was introduced in which the fiber-optic strain sensors were replaced with piezo-strain sensors. The use of clamp-on, piezo-strain sensors provided similar functionality, as the original spool-piece, fiber-optic-strain sonar technology and complexity of the instrument are more user-friendly.

8.5.11 Sonar-Based Flow Measurement System

Sonar-based flow measurements system utilizes an array of sensors, aligned axially along the pipe, to characterize the speed at which naturally occurring, coherent flow structures convert past the sensor array using sonar processing techniques. Since both single and multiphase flows typically exhibit these coherent structures, the methodology is suitable for a wide range of applications. The naturally occurring, self-generated, coherent structures are the present within the turbulent process flow of Newtonian fluids. The time-averaged axial velocity for a the turbulent pipe-flow velocity profile is a function of radial position, from zero at the wall to a maximum at the center line of the pipe. The flow near the wall is characterized by steep velocity gradients and transitions to relatively uniform core flow near the center of the pipe. Naturally occurring, self-generating, turbulent eddies are superimposed over the time-averaged velocity profiles. These coherent structures contain fluctuations with magnitudes on the order of 10% of the mean flow velocity and are carried along with the mean flow. These eddies generated within turbulent boundary layers remain coherent for several pipe diameters and convert at, or near, the volumetrically averaged flow rate in the pipe. Although this description of naturally occurring coherent structures is based on empirical and theoretical understanding of turbulent Newtonian fluids, additional mechanisms serve to generate similar naturally occurring, coherent structures in more general, non-Newtonian multiphase flows that also convert at or near the volumetrically averaged flow velocity.

Sonar technology-based flow meters with coherent structures within pipe-flow sonar-based flow meters use the convection velocity of coherent structures (eddies) within pipe flows to determine volumetric flow rate. The sonar-based algorithms determine the speed of these structures by characterizing both the temporal and spatial frequency characteristics of the flow field. For a series of coherent eddies converting past fixed array of sensors, the temporal and spatial frequency content of pressure fluctuations is related through a dispersion relationship, expressed as follows:

$$V\ k\ \text{convect} = \omega/(1), \tag{8.25}$$

where k is the wave number, defined as $k = 2\pi//\lambda$ in units of 1/length, ω is the temporal frequency in rad/sec, V convert is the convection velocity or phase speed of the disturbance, and λ is the spatial wavelength.

8.5.12 Pressure Pulse Technology

A pressure pulse telemetry system that can be applied to downhole devices for communication in the flowing wells and used to provide a wireless alternative to existing data transfer and actuation methods to find the reservoir boundary.

The telemetry was first applied to a downhole PT gauge, creating the PulseEight Wireless Gauge, which expanded the limited functionality of a memory gauge. This provided a means of adding real-time downhole data from an existing well to reservoir models rather than waiting for the memory gauge to be pulled to surface before analysis could begin.

Since then, the PulseEight device has been modified and upgraded to include PT monitoring, remote valve actuation, and downhole regulation of flow/pressure and has been through an in-depth qualification program. PulseEight system is now a modular monitoring and control device with semi-duplex wireless communication. The device is powered using primary battery cells and can transmit downhole pressure and temperature data to surface as well as act as a remotely operated variable-position flow control valve.

The device is designed to be retrofitted within the production tubing using standard intervention techniques. This allows mature fields and existing wells that have already embraced the digital oil field in their topside infrastructure to install intelligent downhole systems cost-effectively and without a full workover. This enables full reservoir automation for existing assets, reducing the man hours in data analyses/sorting and allowing faster implementation of reservoir optimization techniques.

The PulseEight Wireless Intelligent Completion can be used in several different scenarios, with the possibility of using multiple devices in a single well to provide a range of functionalities, from PT monitoring, interval control, gas hydrate prevention, and downhole regulation to simple on/off plugs that can actuate on command or autonomously based on well conditions. Since the valve can be set anywhere in the production tubing using conventional bridge plugs, there is great flexibility and range in the device's performance objectives.

The service provider had planned on extending battery lifetime at downhole conditions to open up the possibility of "life of well" applications for wireless technology.

8.5.13 Modular-Based Multiphase and Wet Gas Flow Meter

Ultrasonic flow meters use sound waves to determine the velocity of a fluid flowing in a pipe. At no flow conditions, the frequencies of an ultrasonic wave transmitted into a pipe and its reflections from the fluid are the same (Delta-T). The transmitter processes upstream and downstream times to determine the flow rate. *Operators today more than ever seek solutions* for *good-quality data*. One of the widely used modular multiphase wet gas flow came in the market from Roxar, Norway. Images for Roxar MPFM 2600 MVG modular-based multiphase and wet gas flow meter operators today more than ever seek solutions for good-quality multiphase flow measurements directly from the wellhead.

Modular-based multiphase and wet gas flow meter operators work on a good-quality multiphase flow measurements directly from the wellhead. This type of flow meter works on transit time and SRO in order to know non-dehydrated gas flow. Enhance production and efficiency to make marginal fields more viable. The vision of one multiphase flow meter per well brings it closer to global reality.

8.6 MONITORING TECHNOLOGY, DOWNHOLE RESERVOIR SENSORS

IWC Hardware can be broadly divided into the monitors (sensors), analysis (surface computers), and control (ICVs). Sensors used in IWC are installed downhole for the purpose of measuring parameters such as pressure, temperature, flow rate, and seismic waves.

They may be in the form of downhole gauges or optical fibers.

Fiber-optic sensing for distributed temperature measurements has been used in defense, health science, lighting, electrical, and electronic industries before it was brought into the petroleum industry. In the early 2000s, the technology was introduced to the oil and gas industry for downhole temperature and pressure monitoring. Optical fiber consists of a core and a cladding layer, selected for total internal reflection due to the difference in the refractive index between the two.

In practical fibers, the cladding is usually coated with a layer of acrylate polymer or polyimide. This coating protects the fiber from damage but does not contribute to its optical waveguide properties. Individual coated fibers (or fibers formed into ribbons or bundles) then have a tough resin buffer layer or core tube(s) extruded around them to form the cable core. Several layers of protective sheathing, depending on the application, are added to form the cable. Rigid fiber assemblies sometimes put light-absorbing ("dark") glass between the fibers, to prevent light that leaks out of one fiber from entering another. This reduces cross-talk between the fibers or reduces flare in fiber bundle imaging applications. For indoor applications, the jacketed fiber is generally enclosed, with a bundle of flexible fibrous polymer *strength members* like aramid, in a lightweight plastic cover to form a simple cable. Each end of the cable may be terminated with a specialized optical fiber connector to allow it to be easily connected and disconnected from transmitting and receiving equipment. Different types of control cable and power cable are shown in the cross-section of a fiber-optic cable in Figure 8.21.

FIGURE 8.21 Cross-section of a fiber-optic cable.

The importance of sensors in the intelligent well system cannot be overemphasized. Important real-time and future decisions are based on comparisons of readings and measurements that are derived from the sensor to the flowing reservoir model. For example, a downhole pressure sensor's readings during production can be analyzed and compared to the expected geothermal gradient. This can help to detect a zone of abnormal pressure (under-pressure or overpressure), thus preventing a catastrophe. It also enhances the computation of fluid performance in the well and can determine when ICDs should be activated. Temperature sensors are used to detect gas influx (drop in temperature signifies an increase in gas influx (Joule Thomson effect)) and apportion zonal flow rates. The measurements provided by the sensors provide data, which must then be interpreted to provide the required information. Issues with reliability have led to the replacement of electrical systems with fiber-optic systems, which have a higher reliability and higher temperature tolerance capacity capable of efficient operation in difficult environments. Reed et al. have proven that increased reliability of IWC components have led to an evolution of its application from its initial target of well life extension, to include reservoir management.

Downhole data monitoring technology is the "eye" of intelligent well technology. Unlike the traditional method of processing the downhole data through the calculation of the data on the well, the intelligent well monitoring technology is to obtain the downhole real-time data directly. In the 1960s, the first batch of permanent downhole instrumentation (PDG) was formally used.

8.7 ADVANTAGE OF INTELLIGENT WELL COMPLETION

In spite of manifest advantages of intelligent well technology, its development is also challenging. First, the establishment of intelligent well system has the same difficulties as traditional completion technology: small space, large depth, and complex environment. Compared with digital oil and gas well, underground takes millimeter as the basic unit. The oil and gas underground environment measurement is difficult and incomprehensible.

In addition to the difficulties inherent in oil and gas wells, intelligent well technology has its own unique problems: The underground working environment features high temperature, high pressure, corrosion, electromagnetic interference, etc., posing higher requirements for the sensors and actuators placed underground. And again, the components of domestic intelligent well technology are solely imported at present. There is also a high-cost issue to curb the development of intelligent well technology for low-productivity service well.

8.7.1 Limitations of Intelligent Well Technology

8.7.1.1 Input–Output Ratio

In low-productive wells, there may be a marginal comparison between cost and benefit output, which in turn reduces the final economic benefits. Even some wells have an input–output ratio of less than 1, resulting in a loss situation. Therefore, when using the intelligent well technology, the expected production of the well is limited. In addition, the cost of intelligent well technology does not increase linearly; it has a higher investment in the initial stage. Even if the expected production of the well is high and the daily production is insufficient, the investment return period will be prolonged, and the final economic benefit will be reduced. Based on the two considerations above, it is generally believed that the daily production of the well using the intelligent well system should not be lower than 68% of expected planned production.

8.7.1.2 Requirements for Well Conditions

Since the downhole space of the oil and gas well is small, and the intelligent well system has more equipment and facilities, there are certain requirements for the well condition. In the implementation of intelligent well, interlayer packer, inflow control tool, and multiple cables shall be installed in the wellbore. In terms of the packer size of intelligent well system products, the wellbore size is generally not less than 117.8 mm. Intelligent wells are generally suitable for self-drilling wells, gas lift wells, and installation of large displacement pump wells (such as electric submersible pumps). Non-self-injection wells need to be equipped with more wellbore tools, which will lead to further reduction of downhole space, affecting the installation of intelligent well equipment and tools. As a result, the intelligent well system cannot perform its full function or even function properly.

8.8 SOME OF THE IMPORTANT DEVELOPMENT STATUS QUO OF INTELLIGENT WELL TECHNOLOGY

In the 1990s, the concept of intelligent well technology has been implemented in abroad, and at the end of the 20th century, a number of complete sets of intelligent well technology-specific products have been developed and put into production. After decades of development, foreign large-scale oil and gas production equipment enterprises led by Schlumberger, Halliburton, Weatherford and Baker Hughes have further improved the technology of intelligent well from the aspects of electrification, electronics, remote communication, and hierarchical control (Table 8.1).

TABLE 8.1
Intelligent Well Technology Products Developed by Service Providers

Product	Company	Technical Characteristics
In Charge	Schlumberger/ Baker Hughes	Twelve production layers can be controlled independently; single-layer flow can be adjusted sleeplessly; first all-electric cable intelligent well system
Intelligent Zone Compact	Schlumberger	Intelligent Segmentation of Long Horizontal Well; comprehensive integrated intelligent well system
Intelligent Well System	British Shell	Production switching between Lower Ness, Etive-Broom, and Rannoch
In Force	Baker Hughes	Permanent Adaptive Meter; Throttle Valve Over Packer
Smart Well	Halliburton	Realizing Intelligent Production of Three Reservoirs or Branch Well
Intelligent Well System	Roxar	Hydraulic Intelligent Well System for 4-reservoir Control
Intelligent Well System	NHAA	The first multi-fiber sensor smart well in the field of intelligent well
Multi-Node	Baker Hughes	The first all-electronic intelligent well system in the world

8.8.1 Monitoring Technology

The development status quo of intelligent well monitoring technology is shown in Table 8.2.

TABLE 8.2
Intelligent Well Monitoring Technology by Service Providers

Product	Company	Technical Characteristics
Inter ACT	Schlumberger	On-demand query and remote control of key well site parameters; web-based collaborative work mode
Acquire	Halliburton	Distributed Temperature Monitoring System (DTS); data basic processing; providing multi-well management scheme
DACQUS	Roxar	TCP/IP, ODBC, OPC and other interface protocols communicate with third parties
Optical sensing system	Weatherford	Supervisory Control and Data Acquisition (SCADA) is associated with the remote terminal MODBUS protocol
PROVson	Promore	SCADA implements data remote communication; Promore DATA Web network acquires data
Optimize	ABB	Display data trends in real time and offline
ZA-Gauge	Halliburton	Capillary pressure measurement system; pressure sensor with high reliability and long life, but more ground facilities
PDMS	Pioneer	Low cost; wide application; especially in shallow well, normal temperature well, low cost well have obvious advantages
PR300	WTS	Storage underground electronic pressure gauge
Promorey	Promore	Electronic resonance film pressure measuring system; less underground electronic components and strong anti-interference ability

8.8.2 Data Processing Technology

Data processing technology is the "brain" of intelligent technology. The intelligent well monitoring technology continuously acquires data in the well for a long duration. Original data are characterized by massive data set, including abnormal points and noise, and multiple stages of pressure variation. Such data cannot be directly used for reservoir modeling or interpretation with existing well test interpretation theory. In order to effectively utilize the large dataset, appropriate data processing technology is indispensable. The development status quo of data processing technology for intelligent well is shown in Table 8.3.

8.8.3 Application of Intelligent Well Completion System

The successful application of intelligent well technology exists in multi-layer production, artificial lift, water and gas injection, deepwater subsea well, heavy oil exploitation, thin-layer reservoir, multiple completion well, and marginal reservoir development. More than 3,000 production wells have been equipped with complete intelligent well system, and tens of thousands of other production wells have adopted incomplete intelligent

TABLE 8.3
Data Processing Technology of Intelligent Well Service Providers

Researcher	Technical Characteristics
Schlumberger	Integrated Reservoir Simulator (ECLIPSE), Oil Well and Pipeline Modeling Tool (PIPESIM)
Landmark DSP	Combined Well Solver TM, Asset Solver TM, and Asset Lind TM
Athichanagorn	A multi-step method for processing and interpreting PDG long-term pressure data
Ouyang and Kikani	Polytope nonlinear regression method for long-term downhole testing (PDG) data
Soliman	Application of Daubechies wavelet to pressure signal analysis
Olsen	The problem of "automatic" wavelet filtering and compression for real-time reservoir and production data is studied by using real-time wavelet transform
Masahiko and Roland	Another processing method of PDG data
Li	A dynamic model method for diagnosing linear regions in nonlinear systems
Wang Fei	A new method using dynamic deconvolution and corresponding computer coding
Wang and Zheng	Using wavelet transform method to calculate unknown flow through downhole instantaneous pressure data
Liu and Horne	Used nuclear convolution method to interpret pressure and flow data measured by downhole permanent monitoring instruments
Tian and Horne	Deconvolution of pressure data using kernel ridge regression
Geir	Monitoring the reservoir status in the near-well zone by integrating Kalman filtering
Zafari and Reynolds	Prediction of reservoir state change using En KF assimilated production data
Alireza	Established an automatic history matching workflow based on differential evolution algorithm

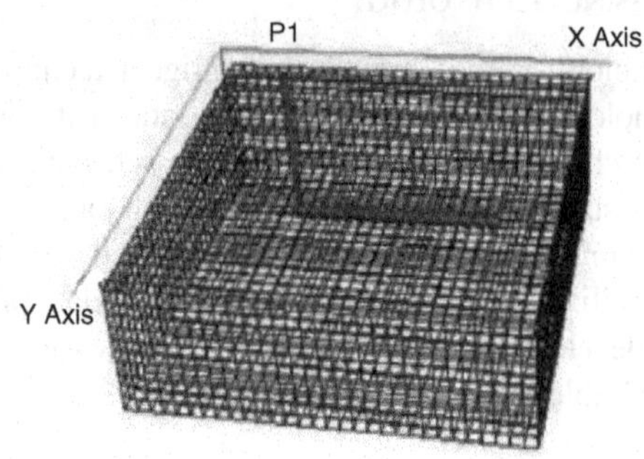

FIGURE 8.22 Multilateral well model.

FIGURE 8.23 Horizontal well model.

well technology such as downhole monitoring technology (pressure, temperature) and downhole flow control technology. Although most intelligent well systems are used at subsea well, the intelligent digitization of land-based oil and gas exploration has become a trend. In the past, intelligent wells were designed with two sets of intelligent completion systems, namely, "direct hydraulic + downhole electronic measuring instrument" and "direct hydraulic + downhole optical fiber measurement." A downhole fluid control valve and a traversing packer are designed, and they can traverse six ¼-inch hydraulic pipelines or optical cables with inner diameter not less than Φ60mm and outer diameter less than Φ156mm. A fiber-optic measurement system with a 3-point horizon and measurable in-column and annulus parameters was designed. Swell packer has to be included here.

In ONGC KG offshore field, a fiber-optic pressure sensor with metal insulation and wavelength demodulation is developed. Laser fusion technology is applied in the probe for stable operation of the pressure sensor in the downhole condition with high temperature and pressure. ONGC had carried out mechanical design covering downhole inflow controller, interlayer packer, wellhead passer, and downhole fluid control valve. And the intelligent well production forecast software was developed to guide and optimize the intelligent well production strategy.

8.8.4 Reservoir Characterization for Intelligent Well Completion

Promt and judicious acquisition, evaluation, and processing of reservoir data are paramount or optimal method of reservoir characterization. Without these, the reservoir is considered a "blind zone" with a high degree of uncertainty, making it difficult for reservoir engineers to manage the reservoir soundly. Data from Intelligent Well downhole sensors are interpreted to enhance and update production models and simulations. This increases the clarity of reservoir connections and reduces volumetric and deliverability uncertainties, where IWC technology is applied for reservoir management and characterization. A production well is fitted with downhole temperature sensors that record a temperature profile during producing life. An analysis and comparison of the measured temperature profile with the geothermal gradient makes it possible to determine the production interval and mass flow input from each reservoir layer. Issues of poor or insufficient reservoir data have embarrassed reservoir engineers in the past, forcing them to reduce estimates of recoverable reserves from oil fields. This situation can be prevented by a combination of maximum reservoir contact (MRC) well and IWC, resulting in a higher and more accurate reservoir description is summarized the positive effect of IWC on reservoir characterization as providing real-time downhole information from the production layers, thus increasing oil recovery through better management.

8.8.5 Overview of Reservoir Engineering Model for IWC

Advancements in hydrocarbon exploration and exploitation technologies have led to upgrades in the complexity of well architecture, and intelligent wells are considered the most advanced method of well completion available. Though the concept of intelligent wells has been in existence for over a decade, it continues to receive a lot of attention, due to the enormous amount of accruable benefits, if effectively applied. Foremost on the list of benefits are the accelerated oil production rate and upsurge in reservoir recovery factor. The functions of IWC are considered two-pronged, i.e., monitoring and control. The downhole sensors act to measure various properties of the wellbore, while the ICDs control and regulate the fluid flow at diverse points based on the results from the sensor. Both the monitoring and control sectors must work in synergy for a well to be truly considered intelligent.

The concept of IWC has been investigated from different perspectives, ranging from its applicability under different conditions and multiple scenarios, to the benefits and challenges encountered in operation. Extensive knowledge sharing on application of IWC technology has bolstered its growth, which was deliberated in the earlier chapter. A review on various reservoir-related aspects has been carried out on previous IWC design and applications, and these are summarized below.

8.8.6 Maximizing Reserves Recovery

Maximizing reserves recovery using horizontal wells requires management of fluid flow through the reservoir. One increasingly popular approach is to use ICDs that delay water and gas encroachment and reduce the amount of bypassed reserves.

With advances in drilling technology over the past 30 years, horizontal and multilateral wells have become a primary design type to develop reservoirs from initial development stage. The need to produce efficiently, economically, and environmentally friendly has promoted the development of extended reach horizontal and multilateral wells that enable greater reservoir contact and lower pressure drawdowns to achieve higher production rates than conventional vertical wells.

8.8.7 Reservoir Modeling for Horizontal Well: Example 1

Here, the main discussion is limited to basic reservoir modeling concepts for selection of horizontal well. This will be better explaining the idea of various types of modeling with examples. One of the tools of choice for the reservoir simulation was a black oil simulator model. When simulating the performance of a three-phase (water, oil, gas) reservoir, the model used was a simple conceptual block with one producer; this model had a role in designing IWC.

The reservoir is almost homogenous with about 30% porosity, and permeability values in the x, y, and z-directions are 100, 100, and 10 md, respectively. The reservoir expert considers a combination drive system with gas cap drives the most dominant. Depletion drive and edge water drive are the other drive mechanisms present.

Two different downhole completion configurations are considered above. These consist of a convention completion with an uncontrolled base case and IWC. Also, two different types of well trajectories were envisaged. These are a horizontal well exemplary configuration, which had a total long length, divided into 16 segments, and a liquid flow rate limit of 3,000 STB/D with the multilateral well flowing at 6,000 STB/D—and installing binary (open/close) ICVs at selected segments to control both gas and water. It had employed two of the reactive control strategies to help achieve the smartness of these wells. One of the configurations is to shut off the well. This also applies to the standardized work combination table (SWCT) for zones of excessive water production. This algorithm is explained in Figure 8.24.

This algorithm can be achieved by the ACTIONS keyword in the SCHEDULE section of the data file (Figure 8.24).

The second type of Rich Communication Services (RCS) had ICVs placed at both water and gas segments, as explained in Algorithm 1 (Figure 8.24).

This strategy was arrived at by shutting off all ICVs with the highest amount of solution GOR and SWCT for both gas and water zones, respectively, after every time step of 45 days. Modeling achieved this by using the ACTION X keyword in the schedule section of the eclipse data file. This RCS was called algorithm.

8.8.8 Field Problem Description: Example 2

MRC is an important parameter that has several benefits such as efficient drainage area, delayed coning or cusping, and increased sweep efficiency. MRC can be achieved using smart wells such as horizontal wells, multilateral wells, and ERD. However, this increased wellbore length has led to some problems in producing from such a well. Higher pressure drawdown around the heel section as a result of frictional pressure drop of fluid flow in the wellbore causes non-uniform fluid influx along the length of the wellbore and higher production rates at the heel. This often

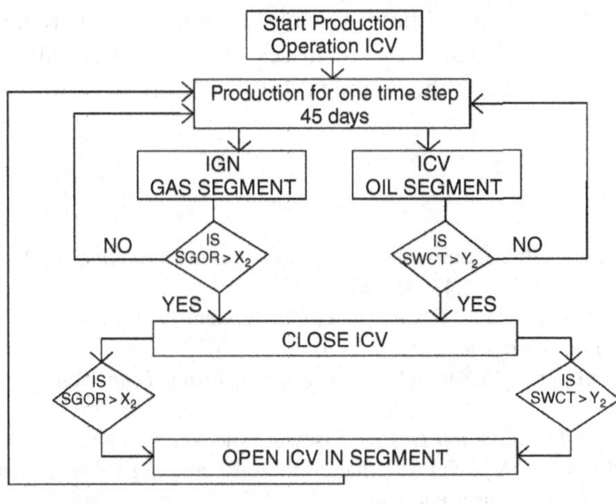

FIGURE 8.24 Schematic of reactive control strategy (Algorithm 1).

leads to early breakthrough of water or gas, which causes a reduction in oil recovery and uneven sweep of the drainage area. With longer contact between reservoirs and wellbore, heterogeneity is more likely encountered, and permeability contrasts along the wellbore can also result in the same phenomena because of unevenly distributed pressure along the wellbore.

A typical drawing of a horizontal well completion with downhole flow control sensors, SCSSV, zonal isolation packer, and annulus bleed valve is shown in Figure 8.25.

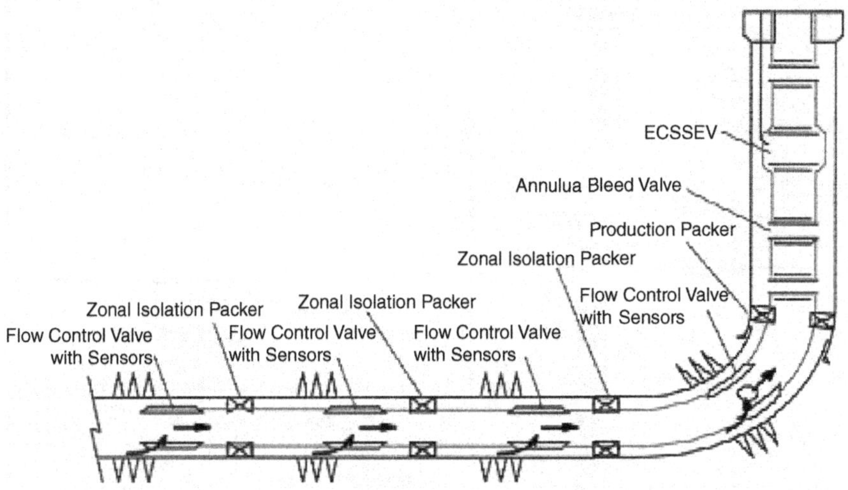

FIGURE 8.25 General principle of well barrier with ICD for horizontal well.

A schematic of horizontal wellbore is given in Figure 8.8, where heel–toe effect is shown. It can be noted that pressure losses along a horizontal wellbore in a homogeneous formation cause the flowing tubing pressure to be lower at the well's heel than at the toe.

BIBLIOGRAPHY

Arumugam, M. Optical Fiber Communication-An Overview. *Pramana*. 2001. 57: 849–869. https://doi.org/10.1007/s12043-001-0003-2.

Arumugam, M., and J. Pagett. "Rotary Steerable System-Overview." *10th International Conference on the Vibration of a Rotary Machine*, 2012.

Azar, Jamal J., and G. Robello Samuel. *Drilling Engineering. Penn Well*, Tulsa, OK. 2007, 237–275.

Bellarby, Jonathan. *Well Completion Design*, Elsevier, The Netherlands. 2009. 642–655.

Cui, Xiaojiang, and Ying Li. A Novel Automatic Inflow-Regulating Valve for Water Control in Horizontal Well. *American Chemical Society*. 2020. 5: 28056–29072.

Mahmood, Md., and Z. B. Sultan. A Smart Well Completion System Review: Route to Smartest Recovery. *Journal of Nature Science and Sustainable Technology*. 2018. 12 (2): 107–118.

Qumar, S. Z., and T. Parvez. Design and Manufacture of Swell Packers: Influence on material Behaviour. *Materials and Manufacturing Processes*. 2012. 27 (7): 721–726.

Zeng, Q., and Z. Wong. A Novel Autonomous Inflow Control Device Design and Its Performance Prediction. *Journal of Petroleum Science and Engineering*. 2015. 126: 35–47.

9 Reservoir Management and iCloud Storage

9.1 THE BASIC CONCEPT OF RESERVOIR MODELING

In recent years, seismic data, especially three-dimensional (3D) seismic, have gained tremendous importance in oil and gas reservoir management and planning. Traditionally, 3D seismic surveys provide valuable reservoir structure type, folding, and faulting details. Improvements in recording and processing techniques, coupled with advancements in processing, interpretation, and visualization software, have revealed the seismic expression of heterogeneities in complex oil and gas reservoirs and even fluids saturation within the rocks.

Conventional seismic data processing makes the simplifying assumption that rocks are composed of horizontal layers; the reflection point in the subsurface is halfway between the seismic source and receiver. Seismic traces can be processed and displayed as a seismic transit time-domain function when this is approximately the case. Paths with a common reflection location but a different source-to-receiver distance are "stacked" to image the rock layers clearly, without creating a detailed model of the rock velocities as they vary both horizontally and vertically.

These assumptions will work reasonably well when slow structural changes occur in the rocks. An additional process called pre-stack time migration is then required to make adjustments in structural positions and shapes.

In complex geologic structures, the assumptions used in conventional processing do not hold. This causes the images focused poorly during stacking process. They look "fuzzy" and dip the position of the reflection images correctly. To solve this problem, geophysicists had developed a method for converting the data from seismic travel time mode to depth domain and doing the migration before the stacking. This pre-stack depth migration requires an iterative procedure of estimating a model of the rock velocities and using it to image the data. Then, the images give a better estimate of the velocity model, which is used to get a better picture. The pre-stack model of depth migration image is sharply focused when the correct model and reflection are always placed correctly. All the data collection of seismic surveys are stored in iCloud for safe storage and used for reservoir modeling work.

Reservoir models are typically of two types:

- Geological models were developed by geoscientists and aim to provide a static description of the reservoir before oil production.
- Reservoir simulation models are created by the reservoir modeler and use the finite difference methods to simulate the flow of fluids within the pool during the field's production life.

DOI: 10.1201/9781003307723-9

A single "shared earth model" is often considered for both purposes. Generally, a geological model is developed at a relatively higher (acceptable) resolution than a reservoir model. A coarser grid for the reservoir simulation model typically relies on a less sophisticated model with approximately two orders fewer cells. Practical values of attributes for the simulation model are then derived from the geological model by various upscaling processes. Similarly, if no geological model exists, the attribute values for a simulation model may be determined by sampling geological maps.

There is much uncertainty in reservoir properties, and it is often scrutinized by developing several types of realizations of the sets of attribute values. A variety of the resulting simulation models can then indicate the associated degree of economic uncertainty.

The word "reservoir characterization" is generally considered part of reservoir modeling activities up to the point when a simulation model is ready to predict and simulate the flow of fluids in the reservoir.

9.2 STATIC MODELING

The particular technique used for 3D reservoir modeling depends on the reservoir type. For layer-cake type reservoirs, deterministic models can often be made directly from the well-to-well correlations, and the rock properties can be interpolated. For labyrinth reservoirs, on the contrary, deterministic modeling is rarely possible without a very close well spacing. A computer design has been developed to generate equally probable 3D models through "probabilistic" modeling techniques. Most system works are done in three stages. Initially, correlatable reservoir bodies are determined and incorporated into the model. An analog database for the relevant genetic type is used where extrapolations beyond control points are necessary. Second, the uncorrelatable bodies were considered, for which the dimensions were also derived from the database.

In contrast, body orientations are taken from borehole imaging logs or estimated based on the general geological model. Characteristic variograms for the thickness distribution of genetic sand body types in different directions relative to their expected trends are also used. Third, especially when the well spacing is rather large, wells do not penetrate smaller or narrower reservoir bodies. Such bodies are added using statistical estimates of their occurrence from the wells. Their position and dimensions are conditioned by geological modeling rules and the analog database. Jigsaw-type reservoirs are also difficult to correlate in the appraisal stage, and probabilistic modeling is required. Later, a large part of the architecture may be determined, and only limited recourse has been taken to probabilistic techniques.

The following are the various types of geological model/geo-cellular models that can be used for the generation of probabilistic modeling techniques:

- Probability and determinism (reservoir model components)
- Static model
- Grid model
- Rock property modeling
- Dynamic model

- Major laws used in reservoir simulation
- Numerical techniques in reservoir simulation
- Scale/upscale
- Pseudo-effective property
- Black oil model/compositional model

The process compares static geological models with seismic models. In mild cases, it can make quantitative correlations, reducing the spread of possible model configurations. The probabilistic models generated are carefully ranked in order of probability by screening them for the presence of geological anomalies. A limited number of 3D models are selected to represent the likely range of variation in the reservoir regarding hydrocarbon volume, connectivity, architecture, and permeability distribution.

A surface-based modeling approach for constructing a reservoir model in which all geological heterogeneity, whether structural, stratigraphic, sediment logical, or diagenetic, impacts the spatial distribution of petro-physical properties modeled in a more discrete volume, bounded by surfaces. The modeled surfaces can be deterministically interpolated between control lines or points or incorporate a stochastic element where control data are sparse. An underlying grid does not constrain models constructed from surfaces; indeed, the model is generated without reference to a grid. The only difference between "geological" and "simulation" models is that the latter incorporates a grid or mesh to allow the numerical solution of the governing flow equations, the architecture of which is driven by the architecture of the modeled surfaces. This approach to gridding (or meshing) is directly compatible with the next generation of unstructured-mesh simulators. It allows the latter's capabilities to be utilized entirely in modeling complex reservoir architectures. A surface-based approach to model construction may facilitate a step-change in reservoir modeling capabilities. Once the requirement to upscale geological models to a structured simulation grid is removed, there is no need to build geological models restricted by grid resolution.

9.2.1 Reservoir Simulation

A 3D seismic survey impacts the original field development plan. With the drilling of new development wells, the added information is used to refine the original interpretation of the data collected. In the passage of time and the data builds, elements of the 3D data that are initially vague make sense. The usefulness of a 3D seismic survey data lasts for the life of the oil and gas reservoir. 3D seismic surveys help identify the reserves that may produce optimally by avoiding drilling a dry hole.

Geostatistical modeling of reservoir heterogeneities play an essential role in creating more accurate reservoir models. It provides a set of spatial data analysis tools as a probabilistic language to share by geoscientists and reservoir engineers and a vehicle for integrating various sources of uncertain information. The geostatistics method help generally the model for the spatial variability of reservoir properties and correlation between a connected model, which can then be used to interpolate a property whose average is critically essential and stochastically simulate for a property

TABLE 9.1
Types of Data and Their Sources

Data	Source
Structure & isopach maps	3D seismic & well logs
Porosity, permeability & fluid saturation	Well logs, cores & correlations
Fluid contacts & formation tops	Well logs
Reservoir pressure & temperature	Well tests
PVT properties	Bottom hole samples & correlations
Relative permeability	Cores & correlations
Production rates & history	Well test & allocation summary

whose extremes are critically important. The following are the significant parameters needed for estimating the values to contour the data by hand or computer algorithm.

9.2.2 Map Statistics

Geostatistics provides a means of interrogating the data to determine how reservoir property varies with distance and direction from any given point in the rock bed. A map that includes these trends can then be created. Net thickness plotting concepts are similar to any properties that can map the standard measure of variation with distance and, optionally, direction. It is called a variogram, or to be precise, a semi-log variogram. The extent of spatial correlation is called a variogram. It determines the variation of net thickness with distance and direction. Variogram can be calculated independent of order, so they depend only on distance. If the charge ranges are N-S, NE-NW, E-W, and NW-SE, then variograms can be plotted in their corresponding directions, and the ellipse (Schlumberger) software can fit them.

9.2.3 Kriging

Kriging is a type of spatial interpolation that uses z-scores to create a least-squares estimate of data and generate a surface model from scattered data points. Krigging definition may be written as "In statistics, originally in geostatistics, kriging or Kriging, also known as Gaussian process regression, is a method of interpolation based on Gaussian process governed by prior covariances. Under suitable assumptions of the prior, kriging gives the best linear unbiased prediction at unsampled location in the map". Once the variogram ellipse has been calculated, it can be used by a contouring algorithm to guide the estimation of the values of net thickness throughout the map. It is known as kriging in reservoir engineering.

9.2.4 Simulation

An alternative to the single map produced by kriging is to generate many maps spanning the range of possible estimates between the bounding lines. It is called simulation. Simulation creates a mathematical model of the mapped property with the same spatial statistics as the actual data. Many possible maps differ from each other but have the same statistics. Some may have to meet the additional condition to honor the measured data points. The process is called conditional simulation in the usual case where the model keeps known data points. Each map, called a realization, is one of many possible estimates. Maps generated by simulation include more fluctuations than kriged maps because they allow for the extremes in the estimated values. None of the maps represents the "most likely" case. Each of the realizations is likely to represent the actual reservoir conditions.

When comparing maps generated by the different methods, the results are as follows:

- Kriging estimates represent a moving average process, smoothing the resulting map.
- The simulation produces a series of more rapidly fluctuating maps, which all have the same general features, but exhibit the range of possibilities.

9.2.5 Geostatistics

Geostatistics provides an opportunity to measure the uncertainty of the maps it produces. Sometimes, kriging map predicts a relatively sizeable net thickness in the area of a shallow well that did not penetrate the reservoir fully. It may propose to deepen the well and complete it in that position. But thickness assumed may need to be checked by other methods. It can also:

- Estimate variation in three dimension.
- Integrate additional measurements as either hard or soft data.

9.2.6 Basic Concept of Reservoir Modeling

The commonly used reservoir performance analysis/reserves evaluation techniques and estimates can be classified as follows, i.e., reservoir engineering techniques for forecasting, reserves estimation, and reservoir behavior prediction:

- Analogs
- Decline curves analysis
- Material balance
- Reservoir simulation
- Reservoir simulation background
- Model purposes
- Model contents vs. complexity
- Reservoir model elements

Reservoir simulation employs a numerical solution of the differential equations, which describe the physics governing the complex nature of the multi-component system, multiphase fluid flow in natural porous rocks in a reservoir, and other modes of fluid flow elsewhere in a production system. The complexity of the physics that governs reservoir fluid flow leads to systems of coupled nonlinear partial differential flow equations that are generally not amenable to conventional analytical methods. As a result, numerical solution techniques are usually used. Reservoir simulation can be extremely computationally expensive, and complex simulations may rely on costly, high-performance computer-based systems to maintain run times within reasonable timeframes. Otherwise, long runtimes associated with lower-performance computer systems may adversely impact user productivity.

9.2.6.1 Recovery Mechanism

Various mathematical models, formulations, discretization methods, and solution strategies are developed and are associated with a grid imposed upon a particular area of interest in a geological formation. Reservoir simulation can be utilized to forecast the production rates from reservoirs and may consider determining appropriate improvements, such as facility modification or drilling of additional wells, that may be implemented to improve production rate, among other uses.

- Mathematical simulation
- Original hydrocarbon in place
- Reserves
- Ultimate recovery
- Original hydrocarbon in place
- Performance under various scenarios

Here are some of the critical reservoir drive mechanisms that will be discussed to explain the process of emplacement of oil and gas from the reservoir.

9.2.7 Reservoir Drives

Although it has long been recognized that pressure energy drives oil to the wellbore from the reservoir boundary, this fact alone was not sufficient to explain how oil is produced and the reasons for the many peculiar production problems that confront during producing life of the field.

Generally, in an oil reservoir, a very complex set of circumstances causes oil to flow through pore channels to the wellbore and be produced. As production starts, pressure drops in the oil adjacent to the wellbore. This pressure drop drives oil from out in the reservoir toward the point of oil withdrawal. Oil, connate water, and rock are all under compression and occupy less space than low pressure. With oil withdrawal, pressure drops in the reservoir, and the oil, water, and rock expand. Expansion of all these materials influences oil production; however, the combined influence is responsible for producing only a relatively small part of oil initially in a reservoir.

Most oil is driven to the wells in natural production by expanding free gas from within or water from outside the reservoir. Gas to furnish energy to displace oil to the wells came from two sources—gas dissolves in oil at high pressure. It is liberated as the reservoir pressure drops. It is gas-free at original reservoir conditions and in the sand above the oil as a gas cap. Water that furnishes energy to displace oil to the wells comes from outside the oil zone or from the water leg, or "aquifer," that occurs in contiguous sand beyond the extremities of the oil zone.

In an oil reservoir, production results from a mechanism that utilizes existing reservoir pressure. This is the drive mechanism. Each reservoir has a specific recovery mechanism that uses the liberation and expansion of dissolved gas and is principally termed a "dissolved gas drive reservoir." One that uses the growth of a cap of free gas over the oil zone is principally termed a "gas cap drive reservoir." One that uses influx and movement of water from outside the particular reservoir layers is termed a "water drive reservoir."

9.2.7.1 Dissolved Gas Drive

Many oil deposits are found in porous rock, either sandstone or limestone, with the porous area of the formation in which the oil is contained surrounded by dense non-permeable rock. Such deposits immediately present evidence against distant migration in the accumulation process. This could result from oil exchange from a shale source bed with original water from an adjacent sandstone bed by capillary pressure forces. Such forces are ever-present in porous rocks when two non-mixable or "immiscible" fluids such as water and oil occur together. An oil deposit of this nature can be considered a container of a fixed volume filled with oil except for the "connate" water that occurs as a microscopic film on the sand grains.

When produced, a reservoir of this physical nature inevitably becomes a dissolved gas drive type of reservoir. The reservoir's name is not derived from the size and condition of the reservoir but from the energy source that produces the oil. This energy is obtained from the light hydrocarbons in the solution in the hydrocarbon liquid mixture in the pool. These light hydrocarbons are liberated from solution as oil is produced, and reservoir pressure drops, forming a gas phase contrasted to the liquid phase in which the rest of the reservoir hydrocarbon materials remain. Gas phases are highly expansible; the gas furnishes energy to push oil to the wellbore as pressure declines. Typical reservoirs having dissolved gas drive at discovery are the Canyon Reef Trend fields in Scurry Country, Texas; Comodoro Rivadavia field, Argentina; Aqua Grade field, Baha, Brazil; Belayim field, Egypt; Joffre field, Canada; Galeki field of Assam-Aaarakan; Kalol Field of Cambay basin and Heera field of Indian Western offshore basin, Mumbai High area.

Dissolved gas drive reservoirs behave characteristically during their producing life. These behavior patterns pertain to changes in oil production rates, pressure, and ratios of gas and oil produced during the reservoir's life.

After drilling a well into a dissolved gas drive type reservoir and production commences, pressure drops to the point where the well penetrates the reservoir. This pressure drop at the well causes fluids to expand from out in the reservoir, driving oil through the tiny pore channels to the well. Pressure in reservoir fluids declines because of the oil withdrawal. Gas evolves from solution and occurs as tiny separate bubbles in individual pore spaces, occupying space vacated by the departure of the oil. At first, bubbles of gas formed by liberation from the liquid solution of the light components do not move because the round bubbles lodge in the small openings between pore spaces. As oil withdrawal continues, further pressure decline occurs, and more free gas is formed. It increases the size of individual bubbles until they enlarge sufficiently to join in a continuous gas thread through pore channels. The gas thus formed in a threadlike state begins to flow.

Gas flows more quickly than oil because it is lighter, less viscous, and does not cling to the surfaces of the pore spaces in the rock. A chain reaction occurs once gas commences flowing; pressure drops faster and lets more significant amounts of gas formed from light hydrocarbons in the liquid. Small additional quantities of oil produced from the reservoir create small further increases in gas space. The gas thus flows much more quickly while oil flows with significantly increasing difficulty. "Gas-oil ratio," or volume of gas flowing compared to the importance of oil flowing, increases until pressure reaches such a low point that both oil and gas flow ceases. Because of depletion in pressure or energy, most fat in the last stages of production must be lifted by pumps from the well. The energy available is sufficient to push the oil to the well but not enough to lift it to the surface. Fat is more challenging to move to the well, and the production rate diminishes during the latter portion of the producing life of the reservoir. The measured producing gas-oil ratio at surface conditions, in cubic feet of gas per barrel of oil in the stock tank, changes throughout the producing life of the reservoir. At first, some gas liberated from the produced oil is held by tiny restrictions in the pore/channels. The produced ratio is less than the ratio of the gas initially dissolved in the oil in the reservoir. The produced gas-oil ratio increases when gas in the pore channels becomes long continuous threads of free gas and starts to flow from the pool. It continues to increase until some low reservoir pressure condition is reached. When this happens, the measured ratio at the surface may diminish because, at lower pressures, the volumes of gas in the reservoir become more nearly equal to the volumes measured at the surface.

Because of its very nature, little or no water is often produced from a dissolved gas drive reservoir; it is a closed trap filled with oil and non-producible connate water.

Recovery of oil from a dissolved gas drive reservoir by its energy is nearly always deficient 5%–15% of original oil being a range of ultimate recoveries encompassing almost all dissolved gas drive reservoirs.

The low recovery from dissolved gas drives reservoirs emphasizes that large quantities of oil remain in the reservoir rocks after they are abandoned. To increase more recoveries from this reservoir type, we must apply more artificial sources.

9.2.7.2 Gas Cap Drive

On many occasions, oil accumulations occurred in which more significant volumes of light materials were present than would dissolve in the oil at temperature and pressure conditions existing in the reservoir. It can be stated that the pressure was not significant enough to retain all the lighter materials in liquid form. The lightweight materials with some intermediate and heavy components formed a free gas phase when this occurred. The free gas bubbled goes to the top of the deposit, where it was trapped, and forms a gas cap over the oil. This excess gas in its compressed state then becomes a source of energy to move oil to the wellbore and lift it to the surface. Suppose a reservoir of this type occurs in an isolated value of rock porosity with a dense rock bed surrounding the reservoir. In that case, the natural energy available to produce the oil comes from expanding the gas-cap gas and expanding the dissolved gas as it is liberated (both occurring with pressure drop caused by oil production). A reservoir of this nature is termed a gas cap drive reservoir. Some reservoirs produced by gas cap drive are the Hawkins field, Wood County, Texas, and the Aga Jari field, Iran.

In the gas cap drive mechanism, oil level in the reservoir drops as production proceeds, and the gas cap expands down into the reservoir section originally containing oil.

Reservoir pressure aims to be higher than in a reservoir with dissolved gas drive. This always depends upon the volume of gas in the gas cap compared to oil volume. The larger the importance of the gas cap, the less the pressure will drop as oil is produced from beneath the lid. This maintenance of pressure on the oil accomplishes several benefits. Dissolved gas is held in solution within the oil itself. The oil is thus lighter and less viscous and will move more quickly toward the wells. A driving action of the expanding gas cap pushes oil down structure ahead of the expanding gas cap, sustaining the production rates of the wells. Gas–oil ratios, however, may increase in wells that are overtaken by the moving gas-cap front. Water production is not a characteristic of a gas cap drive reservoir because such a reservoir does not enclose water except for connate water that may occur as a microscopic film on the sand grains. Because of the effects of gas-cap expansion mechanism on maintaining reservoir pressure and the impact of reduction of liquid column weight as it is produced out the well, gas cap drives reservoirs to tend to flow longer than dissolved gas drive reservoirs depending, of course, upon the quantity of gas in the gas cap and pressure in the pool. Recovery to be expected from such a reservoir will rely upon many things. Size of the gas cap, however, a measure of reservoir energy available to produce the oil, will largely determine recovery percent to be expected. Such recovery usually will be about 20%–40% of original oil in place, but if some other features are present to assist, such as steep angle of dip of formation bed, which allows good amount oil drainage to the bottom of the reservoir, considerably higher recoveries may be obtained. Similarly, a very thin layer of oil columns may limit oil recovery to lower figures regardless of cap size since the entire well stream consists of material originally contained in the gas cap.

9.2.7.3 Water Drive

Today, the most significant natural energy source to produce oil traces its origin back to the ancient seas. The present-day rock formations were formed as sedimentary deposits, the same ancient seas in which present-day oil deposits originated as organic substances. This energy source is the excellent quantity of saltwater existing in the porous channels of rock associated with present-day oil deposits. One must visualize the rock layer occurring over a vast area, with the oil reservoir being a relatively small structural feature into which oil migrated. Therefore, water occurs over a large size compared to the oil in the rock.

Although water is considered incompressible, the total compressed volume is quite large when such significant quantities of total water volume are involved. Even the considerable importance of rock in which the water exists is controlled by water pressure. As oil is produced, the reservoir pressure declines when oil is withdrawn from the reservoir. Water then replaces the oil produced because of the expansion of the minutely compressed water; the reservoir is termed a "water drive reservoir." Many of the most important reservoirs in the world are produced by energy supplied by water drives. The East Texas field, Leduc field, Alberta, Canada; Burgan field, Kuwait; Gela field, Sicily; and the Wafra field, Neutral Zone, Arabia, are typical water drive reservoirs.

The gas-oil ratio remains about the same in a water drive reservoir because high pressure prevents gas from evolving to form a high gas saturation. Water production begins in wells, it starts at a low level in the structure because the water reaches these wells first and displaces the oil. Water production in these wells started to increase until they must abandon them because oil production is deficient in justifying its operation. The energy and action produce a much more significant portion of oil in the reservoir than solution gas drive or gas cap drive. In such a case, recovery usually will be approximately 35%–75% of the original oil in place.

9.3 RESERVOIR HETEROGENEITY

Reservoirs are non-uniform in their respective properties such as permeability, porosity, pore size distribution, wettability, connate water saturation, and fluid properties. These variations can be actual and vertical and are caused by the depositional environments and subsequent events, e.g., Catagenesis.

According to Reed et al., experience has indicated that the profitability of Intelligent Well Completion (IWC) application dramatically depends on the reservoir's inherent, pre-existing properties. The extent to which production is incremented depends on the field properties. The reservoir heterogeneity causes a variation in the deliverability of each layer/zone. The reservoir must be assessed and verified suitable for IWC technology before adopting it. Mr. Ebadi et al. researched the application of IWC to heterogeneous reservoirs.

The permeability and porosity values of the reservoir were varied using the geological properties: Coefficient of Variation (Cv) and Correlation Length (CL). Similar permeability values are grouped; increasing Cv increases the disparity and range of permeability values in the reservoir model. Based on the reservoir models, Mr. Ebadi et al. deduced great potential for added value by applying IWC to a heterogeneous reservoir.

9.4 RESERVOIR SIMULATION AND ICLOUD STORAGE

We do not need to use iCloud Drive; it's an optional feature like all of iCloud services. When it's turned off, all the required documents are stored locally on the device, which means they're always available, even without an Internet connection.

9.4.1 iCloud Backup

It's used for all iPad, iPhone, and iPod backups you've made without using iTunes. iCloud Drive: This is for all your documents (including Mac Desktop and Downloads data) and data from third-party apps on Macs and iOS devices that store data in the cloud. In the cloud platform, high-performance computing (HPC) is designed to incorporate scaling to large numbers of assignments that run in parallel on-demand. Sometimes, this remains a dilemma for many companies in the Oil & Gas industry whether to transition HPC activities to the Cloud or not because of uncertainty. The selection of which HPC applications should migrate to the iCloud is always application-dependent. This section will demonstrate a few case studies assessing migrating reservoir simulation activity to the iCloud. A hybrid cloud solution for HPC is a good kick-off point, as it mitigates a few challenges that entities in the Oil & Gas industry face today. Keywords are reservoir simulation, HPC computing, and cloud computing. The world's eminent technological breakthrough, known as the Fourth Industrial Revolution, digital transformation, and technology, has emerged as the driving core of businesses and organizations. Service companies are continuously striving to incorporate the latest efficient technologies to meet customer needs while reducing the overall operating expense—one of the most popular technologies aiding the transformation toward automation and efficiency in cloud computing. Most industries benefit from cloud technology due to the level of integration that the Internet has brought to operations and transactions. HPC applications are increasingly popular in many fields such as scientific research, academia, business, and data analysis, due to the large gap between scientists' growing computational demands and limited local computing capabilities. Cloud computing service models can be classified into three types:

- Software as a Service (SaaS)
- Platform as a Service (PaaS)
- Infrastructure as a Service (IaaS)

The infrastructure deployment models for cloud computing types can be divided into three main categories: Public, Private, and Hybrid Cloud. It is essential to identify the different kinds of cloud deployment models. The physical infrastructure is controlled by the Cloud Service Provider (CSP) in the public cloud setup.

The Cloud is available to the public, and resources are shared among anyone accessing the Cloud system. The CSP or the organization may own the physical infrastructure in the private cloud setup system. However, the Cloud base storage system is run by a single organization. Any Hybrid Cloud environment is generally a mix of two or more clouds setups. All participating clouds' owners retain the status of a unique entity but share standardized or proprietary technology.

9.4.1.1 iCloud Storage of Data Process

A method for reservoir simulation involves receiving a simulation plan on a specific computer system outside of the cloud computing environment. This plan is prepared by a graphical pre-processor resident. The simulation plan is then executed by implementing a reservoir simulator on multiple compute nodes in the HPC cloud cluster to produce simulation results. Finally, the generated simulation results are sent back to a graphical post-processor on the external computer system.

Method two is where the graphical pre-processor and post-processor are components of an exploration and production software platform resident on the computer system. The computer system may be a desktop computer system, a laptop, or a server computer system.

Method three is where receiving the simulation plan includes receiving at least a part of a model developed by the graphical pre-processor.

In method four, the simulation job generated by the HPC cloud cluster includes encrypted simulation job data, and the method further consists in the HPC cloud cluster: Decrypt the most encrypted simulation job data before executing any simulation job to generate decrypted simulation job data and encrypting simulation output data from the simulation results before returning the generated simulation results such that producing the generated simulation results includes replacing the encrypted simulation result data.

In method five, wherein the HPC cloud cluster considers a storage system within which is stored encrypted simulation job data, decrypting the encrypted simulation job data consists of retrieving the encrypted simulation job data from the storage system. It can keep the decrypted simulation job data in a temporary storage location in the HPC cloud cluster, wherein the simulation results are stored as a temporary storage location by the plurality of particular computing nodes, where encrypting the simulation result data includes retrieving the simulation analysis result data from the temporary storage and storing the encrypted simulation result data in the storage container.

Method six, wherein the storage container is created in response to the computer system external to the cloud computing environment, further comprises deleting the storage container after executing the simulation job and returning the generated simulation results.

Method seven further consists of the central computer system that is external to the cloud computing system: generating a realistic simulation model and the simulation plan using the graphical pre-processor; encrypting the data associated with the simulation plan to develop the encrypted simulation plan data; creating the storage container; storing the encrypted simulation job data in the storage container; detecting and receiving the encrypted simulation output data in the storage container; decrypting the encrypted simulation result data to generate decrypted simulation result data; deleting the storage container; and displaying at least a portion of the encrypted simulation result data.

9.4.1.2 iCloud Computing System

An HPC cloud cluster is provisioned within a cloud computing system; the cluster includes a plurality of compute nodes, including hardware resources from thc cloud computing environment; a reservoir simulator is a resident on the majority of compute nodes. Program code is configured upon execution by at least one processing unit resident in the cloud computing system to receive a simulation job prepared by a graphical pre-processor resident on a computer system that is external to the cloud computing system, initiate execution of the simulation plan by the reservoir simulator to generate simulation output, and return the generated simulation results to a graphical post-processor resident on the computer system that is outside to the cloud computing environment.

9.4.1.3 Data Fabric: the New Paradigm

Data Fabric is an emerging architectural paradigm that enables organizations to centrally monitor, manage, orchestrate, and govern data regardless of where they reside across multiple clouds, on permeability databases, data lakes, data warehouses, or at the edge.

It supports provisioning quality data in the right form, at the right time, to the right consumer for trusted insights and accelerates discovering, cataloging, integrating, and sharing data across the hybrid multi-cloud.

9.4.2 Reservoir Simulation Data Storage Moves to the Cloud

This is another approach related to reservoir Simulation Data Storage in the Cloud system, where one field data provided by Schlumberger is explained. After developing a model, the oil and gas industry performs reservoir simulations to predict oil and gas production. This branch of reservoir engineering is concerned with modeling fluids inside the reservoir for production purposes. Most modern simulators allow for the construction of 3D representations for use in either full-field or single-well models. There are several types of software available with the reservoir simulation service provider, including Schlumberger-Eclipse, Halliburton-NEXUS, Emerson-Roxar, and Rock Flow Dynamics t-Navigator, among many others. There exist several open-source simulators as well, including BOAST and OPM software. Some important visualizations capture in Schlumberger-Eclipse software on cloud storage of reservoir data are presented in Figure 9.1.

Accelerator Core provides a simple way to integrate real-time audio/video into the web application using the OpenTok Platform (www.vonage.com/communications-apis/video). The industry-standard simulators operate with the acceleration value of below 20 (Figure 9.2).

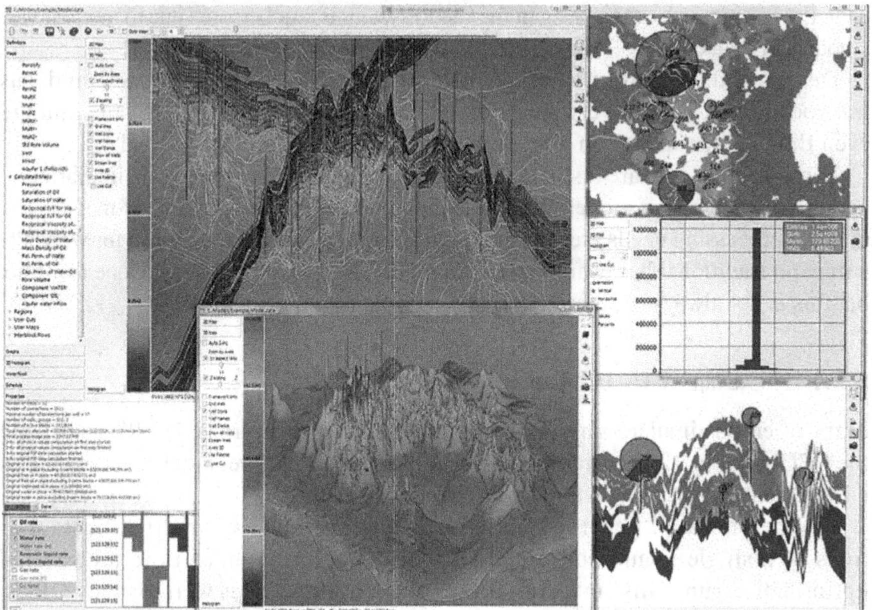

FIGURE 9.1 Schlumberger-Eclipse software on cloud storage of reservoir data.

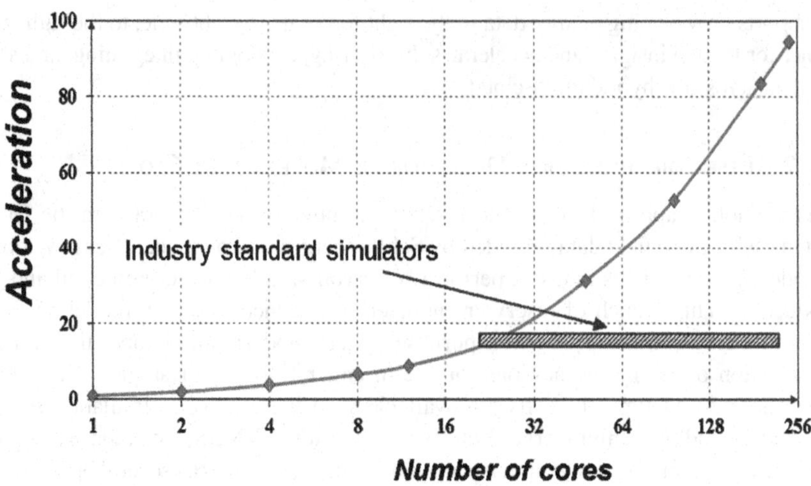

FIGURE 9.2 Core data vs. acceleration value.

9.5 A MODERN RESERVOIR SIMULATOR INTERFACE: CASE STUDY-1

Reservoir engineering is a multi-discipline that includes seismic data acquisition, data processing, interpretation, reservoir analysis, simulation, well and drilling process modeling, and fluid flow modeling. Seismic acquisition and modeling are highly data-intensive—an oil field model in the 3D space can accommodate tens of terabytes of raw field data. Processing such data storing in the Cloud system is cumbersome due to moving vast amounts of data on and off the Cloud.

Design and drilling modeling, including pipe flow modeling (oil and gas transportation), is relatively small in data and processing needs. It can simulate even the big-size model on a modern laptop in a fraction of a second.

Large seismic data are distilled to a relatively small size in reservoir simulation to comprise the reservoir model. While the inputs are in the scale of tens of megabytes, the data processing needs are enormous, with some models that may run for weeks at a time on traditional hardware that needs more CPU power that can reduce runtime, but this is not always straightforward.

9.5.1 Scalability Challenges

Many reservoir simulators have a limitation that can handle/process well data beyond the CPU cores of 12–32 for a single simulation job. Therefore, moving these data to a large cluster or the Cloud may not be beneficial.

On the contrary, general simulation modeling involves multiple instantiations of a single simulation process using different parameters in an assisted optimization run. This embarrassingly parallel setup scales well to many CPUs, and most simulators can take advantage of the speedup. It is called the sweet spot for a cloud platform.

t-Navigator + Rescale = Match Made in Cloud t-Navigator by Rock Flow Dynamics (RFD) is considered an extremely scalable, parallel, interactive reservoir simulator. It has also been shown to achieve near linear speedup using thousands of CPU cores, even on a single instantiation.

Figure 9.3 shows t-Navigator's scalability compared to industry-standard simulator t-Navigator's scalability versus that of industry-standard simulators coupled with an on-demand, pay-per-use cloud solution, such as rescale, oil, and gas business providers will take advantage of a fully scalable solution. Users can spend for the exact amount of computation needed to utilize, as both rescale and t-Navigator offer an hourly estimate. This arrangement allows for 100% resource utilization and permits large clusters to be created on demand to speed up the computations.

Rescale and RFD have been implemented a cloud-based solution for t-Navigator. Users generally upload the input data via encrypted channels, choose the number of CPUs to be used, and press Submit. The simulation results may be seen remotely at runtime from a consumer's terminal. The data are 100% secured when running simulations on Rescales platform.

9.5.2 Case Study-2

To select real-world three-phase models with 39 wells, ten years of historical data must be tested to assess the system's scalability. This model includes 21.8 million active grid blocks and is supported by a powerful 512-node cluster. Each cluster node contains two four-core Intel Xeon 5570 processors in Nehalem, totaling 4,096 simulation cores.

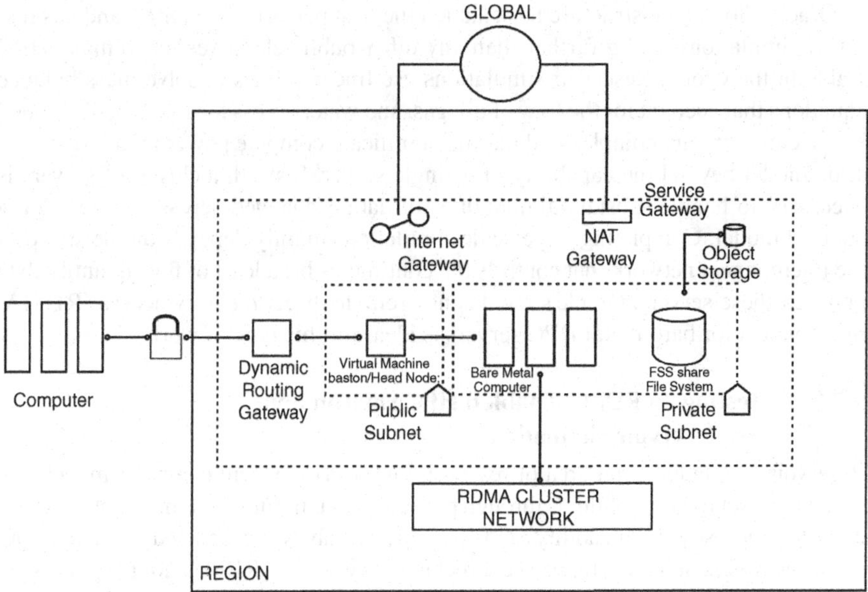

FIGURE 9.3 Architecture of an RDMA-enabled HPC environment on oracle cloud infrastructure.

This type of setup resulted in a computation time reduction from 2.5 weeks down to 19 minutes. The resulting speedup coefficient compared to one calculation core is equal to 1,328.

9.5.2.1 Rock Flow Dynamics

RFD was first introduced in 2005 by mathematicians and physicists, is backed by Intel Capital, and has customers on all continents, including some big names in the oil and gas industry such as Occidental, BG Group, Marathon Oil, Tullow Oil, Petro-fac, Murphy Oil, Penn-West, Tatweer Petroleum, Petro-China, etc. It introduced RFD with a clear path to provide reservoir engineers a dynamic reservoir simulation technology with new state-of-the-art technology that meets the most modern expectations for raw performance, rich modeling functionality, and advanced graphical user interface capabilities.

9.5.2.2 Rescaling

Rescaling is a process to secure a cloud-based, HPC platform for scientific and engineering simulations. This platform allows engineers and scientists to quickly build, compute, and analyze extensive simulations processes easily on demand. Rescaling partners with industry-leading software services to provide instant access to various simulation packages while offering customizable HPC hardware.

Industry professionals use simulation techniques to improve accuracy to make the best possible decisions about drilling a well. Sometimes, the advanced reservoir simulation plays a pivotal role in enhancing decision-making by providing estimated oil and gas production from an oilfield before drilling the wells.

Oracle Cloud Infrastructure has enabled the best performance for oil and gas reservoir simulations, going further than any other public cloud vendor to make HPC viable in the Cloud. Reservoir simulations use linear solvers to solve mass balance equations that account for the flow of oil, gas, and water across locations in the model. These equations are complex and require significant compute power for a large field-scale model beyond the capability of a single server. Instead, a cluster of servers is necessary to handle substantial quantities of data exchanged across a large, tightly coupled model. Completing large-scale simulations mainly depends on the speed of the interconnect network that controls the continuous broadcast of flow quantity data between these servers. Oracle's low-latency remote direct memory access (RDMA) interconnect for bare-metal HPC servers is ideal for this type of work.

9.5.2.3 Testing an RDMA-Enabled HPC Environment for Reservoir Simulation

Reservoir simulation is an area of reservoir engineering where computer models are used to predict the flow of fluids through porous media. In EnKF method, static parameters like porosity, permeability, and dynamic variables are updated to match with real-time production data. In remote direct memory access (RDMA) for reservoir simulation data storage, high-performance computer (HPC) is essential. To demonstrate the capabilities of our HPC infrastructure to support the oil and gas reservoir simulation, we used a transparent and repeatable methodology. With commercial and open-source reservoir simulators available, the open-source simulator Open Porous Media

Flow (OPM) was chosen. We used the Norne reservoir model for the benchmarking tests, including data from a Norwegian oil field part of the OPM Flow datasets.

9.5.2.4 Case Study-3

Society of Petroleum Engineers (SPE), USA, had Comparative Solution Project SPE5 model for developing an 11.025 million cell reservoir model. This model contained one oil producer, one water injector well, and roughly 100 times more total grid cells than the Norne model. It had performed simulations using OPM Flow on Oracle Cloud Infrastructure bare-metal HPC instances with RDMA cluster networking.

Bare-metal BM.HPC2.36 shapes feature 36 Intel 6154 cores with a clock speed of 3.7 GHz, 6.4-TB local NVME drive, and 384 GB of memory. For the tightly coupled reservoir simulation systems, ultra-low-latency cluster networking is crucial. The HPC shapes give users the world's first public cloud bare-metal RDMA network, enabled by a Mellanox 100 Gbps network card. Oracle Cloud Infrastructure uses RDMA over Converged Ethernet (RoCEv2), which delivers 1.5-µs single-hop latency—unparalleled in the public clouds. A single RDMA cluster on Oracle Cloud can scale up to 20,000 cores. The bare-metal computes and networking approach uses no hypervisor or server agents and allows no resource oversubscription. This combination eliminates OS and network jitter, delivering consistent, predictable high performance.

9.5.3 Linear Scaling for Reservoir Simulations

The benchmarking simulations of the modified SPE-5 model demonstrate strong scaling for reservoir simulation on HPC clusters. When a workload increases performance equal to the number of processors, it's called linear scaling compared to the version on a single processor. Figure 9.2 shows the scaling of a model with 11.025 million cells, along with the theoretical linear scaling offered by the gray dotted line. The HPC infrastructure can achieve close to linear scaling for reservoir simulations using simulation models with tens of millions of grid cells. The bare-metal computes, and cluster network offered by Oracle Cloud make this result possible.

Reservoir simulation data are sometimes stored on an Oracle Cloud Infrastructure and other unknown public cloud domains. The benchmarks for the Norne model performed on the bare-metal HPC cluster with the publicly available standards from another public cloud that uses virtualized RDMA.

Bare-metal computes and RDMA cluster networks for reservoir simulation workloads. Oracle Cloud Infrastructure provides better performance for the entire benchmarking range and is almost 25% faster than the public Cloud with virtualized RDMA for the simulations run on multiple cluster nodes. These benchmarks demonstrate a significant advantage offered by Oracle Cloud Infrastructure.

High-performance cloud computing (HPCC), a mature oil and gas industry, is not a new technology but a new method of delivering two new resource components: storage capacity and computational power. Cloud computing is very immature within the upstream oil and gas industry, as the industry has constantly been challenged by storage and computational capability. However, a recent example is considering using an HPCC due to the assurance of several gains, such as flexibility, accessibility, and cost reduction (pay-per-use). New HPCC systems can also be categorized into private, public, or hybrid models. Scenario-based modeling, the probabilistic approach for uncertainty analysis, risk quantification for existing brownfield types of reservoirs, and new perspective green-field reservoirs require a significant amount of computational power and storage capabilities.

9.5.3.1 Open Porous Media Software

OPM initiative was started in 2009 to encourage open innovation and reproducible research on reservoir modeling and simulation of porous media processes. OPM was formed between Equinor (formerly Statoil), SINTEF, the University of Stuttgart, and Bergen, Norway, but several other groups and individuals have joined and contributed over time. The OPM suite of software is mainly developed by SINTEF, NORCE (formerly IRIS), Equinor, Ceetron Solutions, Poware Software Solutions, and HPC-Simulation-Software & Services.

The opening vision was to create long-lasting, efficient, intelligent, and well-maintained, open-source software for simulating fluid flow and transport in the porous media. The scope has later been extended to provide open data sets, thus making it easier to benchmark, compare, and test different mathematical models, computational methods, and software implementations. OPM simulator flow logic explains the process of upscaling the model shown in Figure 9.4.

OPM Flow is a reservoir simulator that includes an upscaling tool and a selection of tools to support experimental software pieces. The corresponding source code is divided into six modules, as shown in Figure 9.4, and is organized as several git repositories hosted on github.com/OPM. The data repositories, OPM data, and OPM tests are also hosted there and contain example cases and data used for integration testing.

9.5.3.2 Performance Compared with Other Clouds

A pilot project was where Amazon Web Services was used as a public HPCC, with an industry-standard reservoir simulation software known as Eclipse has been used. The project conceptualized designing and developing a secure, lean-agile technology model, software vendor agnostic. It is believed that this would drive efficiencies and reliability by dynamically scaling up or down computing clusters depending on needs. This part presents a recommended development methodology for creating tactical and strategic roadmaps for leveraging trends in HPCC to further unlock potential in the industry by driving innovation and business value.

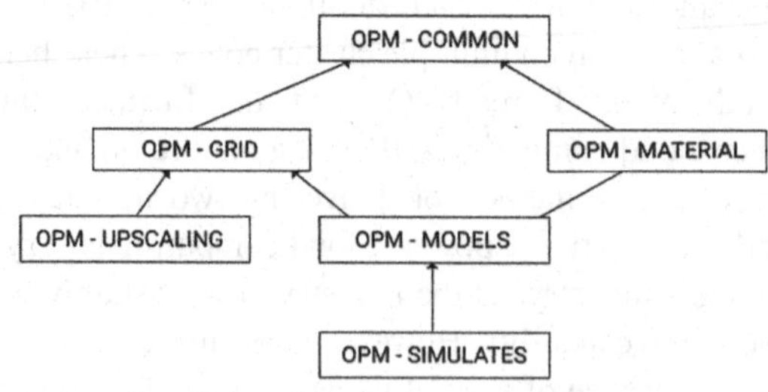

FIGURE 9.4 OPM source model.

9.6 A DELIBERATION ABOUT SCHLUMBERGER'S NEW DELFI COGNITIVE E&P ENVIRONMENT ENABLING FASTER TURNAROUND IN THE EXPLORATION-TO-PRODUCTION LIFE CYCLE

The INTERSECT high-resolution reservoir simulator of Schlumberger had designed many reservoir challenges. Operators can simulate detailed geological features and quantify uncertainties by creating real oil and gas production scenarios. With the integration of precise models of the surface facilities and field operations, the INTERSECT simulator produces reliable results constantly updated in real-time data exchanges.

The main benefits of using INTERSECT simulator are as follows:

- High-resolution modeling for complex geological structures
- Well-completion configurations for complex wells
- Detailed chemical-enhanced-oil-recovery (EOR) formulations method
- Application of steam injection and other thermal EOR methods
- Advanced production controls in terms of reservoir coupling and flexible field management
- Flexibility to script customized solutions for better modeling and field management control

9.6.1 A Secure, Cloud-Based Environment

DELFI is a secure, cloud-based space built on critical premises, making it unique in the Schlumberger software portfolio. It generally harnesses data, scientific knowledge, and domain expertise in every part of the E&P value chain. The DELFI makes applications and workflows accessible to all users and enables team members to build shared workspaces for data, models, and interpretations while respecting proprietary information boundaries.

9.6.2 Secure

It also safeguards customers' data, which have been a Schlumberger priority for many years, so the DELFI system is fully compliant with all software regulations. It is also totally supported by cloud vendor security, operational procedures, and audits.

9.6.3 Collaborative

The DELFI environment provides a shared space for teams to work together, independent of role, workflow, or physical location. The result is a borderless collaboration where all stakeholders benefit directly from each other's expertise and insights.

9.6.4 Open

Openness and extensibility environment make it straightforward to integrate, share, and build on all these available data sources, workflows, intellectual property, and applications. By making the solutions in the DELFI system open and accessible via APIs, extensibility enables proper integration where partnerships and collaboration are taken to a new level. Extensibility has long been recognized as a highly desirable trait of hypermedia systems. Furthermore, many hypermedia systems claim to provide some form of extensibility. However, consensus on what functionality of hypermedia should be extensible, how this extensibility should be presented to extension implementers, and even a rigorous definition of the term itself has evaded the hypermedia field.

9.6.5 Cognitive Technology

The DELFI system puts the full spectrum of available cognitive technologies to work for customers, from artificial intelligence (AI) to analytics. Robust cognitive systems recognize each user to deliver a uniquely personalized experience. Intelligently searching, automating tasks, proactively learning, and enabling customers to predict, prioritize, and advise.

9.6.6 Live Project

New data are automatically shared between tasks across the DELFI environment; each live project is dynamically optimized with the latest information. The entire team can work concurrently on the same project, with the confidence that the information customers are using is always current and accurate. Duplication of work is reduced, team productivity and efficiency are increased, and business continuity is secured.

AI and cloud computing, along with machine learning (ML) and increased automation, are starting to drive a step-change in how we make faster decisions in the E&P industry with higher levels of confidence. These new digital technologies yield greater and much quicker insights in our subsurface workflows, but at cost and without compromising high standards.

One of the primary challenges with exploration and development in conventional settings is the long turnaround times associated with finding and extracting the oil and gas. If service provider can significantly reduce those turnaround times, then traditional settings can become far more competitive when compared with unconventional and non-oil and gas industry-related investments. Several examples show how digital technologies can substantially improve conventional resources' total turnaround time and competitiveness.

Early challenges in the exploration workflow are finding, integrating, and visualizing data. Overcoming data challenges is crucial to success. Cloud computing, modern scalable and flexible cloud data storage, and intelligent data correlations and access enable the integration of substantial data objects at any scale and can visualize those on any device. Such a modern cloud data ecosystem also enables automated unstructured data mining, where cloud and AI technologies can deliver substantial value. Application of natural language processing and computer vision to automatically extract and integrate historical documents into subsurface workflows can provide faster and deeper insights.

Schlumberger has addressed bottlenecks in seismic data processing and interpretation in the geophysics domain, resulting in significantly reduced turnaround times. Integrated cloud platform services allow access to almost unlimited computing power to tackle computationally intensive tasks, resulting in substantially faster end-to-end processing and interpretation workflows. Many steps can be optimized through ML and analytics, which offer insight from the rich history of past decisions by experienced geoscientists. In seismic data interpretation, deep convolutional neural networks are achieving significant successes in the areas of seismic horizons (e.g., top of salt), faults, facies, zones, and geobody, where the techniques are capable of training and predicting across seismic cubes in a matter of hours, generating high-quality picks with very few, if any, false positives. AI also enables the automatic extraction and modeling of structural frameworks in orders of magnitude less time than traditional manual approaches.

The wellbore domain has addressed typical challenges driving insights into exploration opportunities. For example, well logs are obtained from various vintages and typically have missing/corrupted sections, limiting understanding. Schlumberger has developed numerous ML techniques that can be trained on just a few wells and predicted across many wells throughout a basin. These processes have been applied successfully in many basins around the globe, both in conventional and unconventional settings, and benchmarks have shown that they compare favorably with typical processing/interpretation workflows.

Mass automation of subsurface workflows orchestrated by AI enables the generation of many realistic realizations of the reservoir. Cloud computing and AI then become vital in mining the data and highlighting the critical controls and insights that control the economics of the system.

Overall, it has been seen that new digital technologies can provide substantial value across E&P subsurface domains. Developing and applying these new technologies are not trivial, but focused multidisciplinary teams can obtain significant benefits across many areas. We have seen that automation and machine-assisted and augmented interpretation digital systems, powered by cloud infrastructures, deliver significantly faster performance and drive better outcomes and more agile business behaviors.

9.6.7 Synopsis of Use Cases of the Valuable Benefits to Various Customers Using Schlumberger's Digital Technologies (DELFI) Software

9.6.7.1 Woodside Energy

Woodside Energy uses the high-quality field development planning made available to oil and gas companies through the FD-Plan solution in the DELFI cognitive E&P environment from Schlumberger. FD-Plan in the Woodside field is being used with its potential to reduce end-to-end forecasting time by 70% with instantaneous economics for a wide range of subsurface development options.

9.6.7.2 Schlumberger Had Developed a DELFI Environment for Woodside Energy

DELFI cognitive E&P environment had collaborated with Woodside energy; Woodside will leverage this secure cloud-based software environment to increase consistency, reduce study circle time, and foster innovation in its subsurface characterization and development activities.

9.6.7.3 Explore Plan of Equinor Oil Company

Equinor oil company has been using Explore Plan as a one-stop-shop solution for exploration in the capacity of a hydrocarbon exploration management tool that plans and tracks. It facilitates the creation, iteration, and maturation of new and existing opportunities in the exploration funnel to save huge money.

Integrated Schlumberger drilling software technology had helped them reduce by 29% of drilling time. A challenging and abrasive stringer at the base of the 12¼ inch section caused a high level of shock and vibration damage to drilling tools. This damage—combined with severer friction, high dogleg requirements, twist offs, and fluctuating torque values when drilling through interbedded chalk and limestone—caused multiple trips to replace.

Equinor did use the first to mitigate tool damage. Equinor used an active vibration dampener on conventional tools to improve the performance, but it was insufficient.

9.6.7.4 Equinor and Schlumberger Digital Program

Equinor and Schlumberger went into a digital exploration collaboration program wherein it was made possible the fact extraction from unstructured data. Equinor invested significant investment in digitization—securing a global leadership position, and part of the strategy was to join Schlumberger in a collaborative program using DFIFI. Technology Workshop was organized to identify the potential focused area, among which extracting facts from unstructured data were considered a high possible project with number one priority. This is because today E&P industry involves dealing with diverse data from a massive amount of unstructured data from the myriad source; companies have to extract, analyze, classify, and summarize data to fast-track a review of all available data and knowledge related to the E&P asset and quickly evaluate the potential of oil and gas opportunity of interest.

9.6.7.5 Drill Plan Solution's Digital Drilling Enablement and Automation Benefiting OMV's Well Construction Process

The use of full automation allows us to free up valuable time for Schlumberger engineers by eliminating repetitive tasks like building offset databases or writing programs. It improves the efficiency of an integrated project team by allowing everyone from operations, geology, and petrophysics to the completions engineer to work on the same platform, with everyone having the newest data and insights available in real time.

9.9.7.6 Schlumberger and OMV Collaboration

Schlumberger and OMV had developed new AI & Digital solutions enabled by the cloud base DELFI cognitive environment across OMV's global operation. They had jointly innovated on projects to support specific business goals like reducing well planning times and accelerating the field development planning, supported by leading digital technologies deployed within the DELFI environment.

9.6.7.7 Redefining Seismic Interpretation: 80% Time Reduction—A Use Case by INPEX

Using machine learning (ML) to identify and interpret faults and integrate them into the reservoir modeling process enhances reservoir development. The ML model's fault prediction has produced fairly accurate results in predicting the overall structures in the input seismic cube. This type of fault prediction provides a basis for initial fault extraction and structural modeling, highlighting areas where the prediction result could improve.

Extending the petition to a user-driven ML model with user-defined fault labels yielded a massive increase in the accuracy of the fault forecast. The geophysicist endorsed this set of faults and used it to convey a second structural model realization. Using an increased approach to fault identification significantly improved the domain-science-driven workflows and delivered substantial value to the interpretation loop—achieved an average 80% time reduction in the overall structural interpretation for this project.

9.6.7.8 Smart Surveillance: Automatically Interprets Field Data 24/7

The combination of high-frequency records and occasional data, along with their respective time and date, allows for the application of data analytics and machine learning solutions. The Supervisory Control and Data Acquisition (SCADA) system provides necessary data about operational conditions during any occurrence, including wellhead temperatures and pressure, as well as well status and valve positions. This data enables AI to measure the feedback from the well during specific operational processes. AI solutions can apply pattern recognition, machine learning, and data-driven analytics to typical oil and gas production operations, particularly for naturally flowing oil wells.

These cases include:

1. Automated identification of the reservoir environment (fractures, matrix, or dual porosity) based on bottom hole or wellhead pressure tendencies, which can identify radial or linear transient flow patterns.
2. Diagnosis of water-cut instabilities based on pressure drop changes through the wellbore, altering the wellhead pressure declination rate.
3. Identification of hydrate formation based on surveillance of flow line pressure variations and the identification of uncelebrated choke valves in the surface equipment.

Thus, the solution automatically interprets field data (24/7) and provides valuable information to the technical team—it has become a helpful assistant for petroleum engineers. The solution enhances the supervision and interpretation levels, which has led to increases in efficiency, improved Health Safety & Environment (HSE) levels, and several economic benefits.

Simultaneous oil production and water injection: With dual concentric tubing completion, ESP technology has improved recovery factor without drilling a new well.

9.6.7.9 Petro-Hunt

The one drilling well plan of Petro-HUNT will be discussed in this section. Petro-Hunt's well planning group was able to improve multi-discipline collaboration between engineering and geology, and interaction with service companies increases planning efficiency by integrating well-planning workflows with a stage-gate approval process to explore more planning scenarios with minimal overhead cost, increase the reliability of the design process with traceability and versioning use automated tools, such as automatic trajectory and anti-collision scanning, that incorporate offset well trajectory data to decrease time, cost, and risk, and enhance innovation in well design and efficiency because of automated workflows.

As a result of the successful application of the customer trial, Petro-Hunt began using the Drill Plan solution to plan wells across its portfolio started in February 2018.

9.6.7.10 Dallas-Based Petro-Hunt Company

Dallas-based Petro-Hunt company improves the drilling plan efficiency by 50% in Williston Basin like drilling risk mitigation and well design mitigation by application of DELFI.

Petro-Hunt did recognize that their Liquid Level control (LLC) operation in the North Dakota field needs more detailed planning in challenging deviate High-Pressure High Temperature (HPHT) well. They were continuously exploring the possibility of enhancing well-designing innovation and increasing planning efficiency to drill and complete the well faster.

9.7 ROLE OF CLOUD BUSINESS

As exemplified by the Amazon Elastic Compute Cloud (EC2), the emerging cloud computing paradigm represents an excellent conceptual foundation for hosting and deployment of web-based services while theoretically relieving service providers from the responsibility of providing the computational resources needed to support these services. Cloud computing system offers multiple advantages: it allows individuals or companies with the market domain expertise to build and run their Software as a Service (SaaS) company with minimal effort in software development and without managing any hardware operations. It helps reduce software complexity and costs, expedite time-to-market, and enhance the accessibility of consumers. While cloud computing holds enormous promise for the future of service computing, several inherent deficiencies in current offerings can be pointed out.

9.7.1 Inherently Limited Scalability of Single-Provider Clouds

Although most infrastructure cloud providers today claim infinite scalability; in reality, it is reasonable to assume that even the most prominent players may start facing scalability problems as the cloud computing usage rate increases. In the long term, scalability problems may expect to aggravate as cloud providers serve an increasing number of online services, each accessed by massive amounts of global users at all times.

9.7.2 Lack of Interoperability among Cloud Providers

Contemporary cloud technologies have not been designed with interoperability in mind. This results in an inability to scale through business partnerships across cloud providers. In addition, it prevents small and medium cloud infrastructure providers from entering the cloud provisioning market. Overall, this stifles competition and locks consumers to a single vendor.

9.7.3 No Built-In Business Service Management Support

Business Service Management (BSM) is a management strategy that allows businesses to align their IT management with high-level business goals. The critical aspect of BSM is Service Level Agreement (SLA) management. Current cloud computing solutions are not designed to support the BSM practices well established in the daily management of the enterprise IT departments. As a result, enterprises looking at transforming their IT operations to cloud-based technologies face a non-incremental and potentially disruptive step.

9.8 THE RESERVOIR APPROACH

The reservoir vision is to enable on-demand IT services at competitive costs without requiring significant capital investment in infrastructure development. This model is inspired by a strong desire to liken IT services to the delivery of standard utilities. For example, a common scenario in the electric grid is for one facility to dynamically acquire electricity from a neighboring facility to meet a spike in demand. It seems that, similar to other industries, where no single provider serves all customers at all times, next-generation cloud computing infrastructure should support a model where multiple independent providers can cooperate seamlessly to maximize their benefit.

More specifically, it believes that to fulfill the promise of cloud computing. Indeed, there should be technological capabilities to federate disparate data centers, including those owned by separate organizations. Only through federation and interoperability can infrastructure providers take advantage of their aggregated capabilities to provide a seemingly infinite service computing utility. Informally, it refers to the infrastructure that supports this paradigm as a federated Cloud. A significant additional advantage offered by the federated cloud approach is that it democratizes the supply side of cloud computing and allows small and medium-sized businesses and new entrants to become cloud providers. This encourages competition and innovation.

As discussed above, one of the limiting factors in current cloud computing offerings is the lack of support for BSM, specifically for business-aligned SLA management. While it can enhance specific cloud computing with some aspects of BSM, complex services across a federated network of possibly disparate data centers are a challenging yet unsolved problem. Service may consist of numerous distributed resources, including computing, storage, and network elements. Provisioning such a service consumes physical resources but should not cause an SLA violation of any other running application with a probability more significant than some predefined threshold. As SLAs serve as risk mitigation mechanisms, this threshold represents the risk that a cloud provider and the cloud customer are willing to accept. With BSM, applications are correctly dimensioned, and their non-functional characteristics (e.g., performance, availability, and security), governed by SLAs, are ensured with the optimal cost through continuous IT optimization. It argues that due to the immense scale envisioned by cloud computing, support for BSM should be maximally automated and embedded into the cloud infrastructure.

In the reservoir model, each infrastructure provider is an independent business with its own business goals. Based on local preferences, a provider federates with other providers (i.e., other RESERVOIR sites). The IT management at a specific RESERVOIR site is fully autonomous and governed by policies aligned with the site's business goals. Once initially provisioned, it may move resources and services to other reservoir sites based on the economics, performance, or availability considerations to optimize this alignment.

Various research addresses those issues and seeks to minimize the barriers to delivering services as utilities with guaranteed service levels and proper risk mitigation.

Cloud computing is the latest incarnation of a general-purpose public computing utility. In recent years, it had seen many efforts toward computing as a utility—from grid computing, which made significant progress in the area of high-performance scientific computing, to attempts at building enterprise-level utilities. None of these attempts will materialize into a general-purpose compute utility accessible by anyone, at any time, from anywhere. What makes cloud computing different is that industry trends such as the ubiquity of broadband networks, fast penetration of virtualization technology for x86-based servers, and the adoption of Software as a Service finally create an opportunity and a need for global computing utility. The reluctance to use online services as a replacement for traditional software is minimizing:

- the success of companies such as salesforce.com proves that with the right set of security warranties and a competitive price, companies are willing to trust even their most valuable data.
- customer relations.
- to an online service provider. At the same time, virtualization has made it possible to decouple the functionality of a system as it is captured by the software stack (OS, middleware, application, configuration, and data) from the physical computational resources on which it executes. This, in turn, enables a new model of online computing.

- instead of specially crafted online software, it can now be thought in terms of general-purpose online virtual machines that can do anything. Finally, as virtualization enters the mainstream, the era of a general-purpose compute utility is now within reach.

BIBLIOGRAPHY

Craft, B. C., and M. Hawkins. *Applied Petroleum Reservoir Engineering*, Prentice Hall, Hoboken, NJ. 1991. 142-282, 273–333.

Dake, L. P. *Fundamental of Reservoir Engineering*, Elsevier Science BV, Amsterdam. January 1983. 191–278.

Hohn, Michel E., and R. Mohan Srivastava. *Introduction to Applied Geostatistics*, Oxford University Press, New York. 1991. 34–75. ISBN 0-19-505012-6.

Sarkodie, Kwame, S. A. Afari, and W. N. Aggrey. Intelligent Well Technology-Dealing with Gas Coning Problems in Production Wells. *International Journal of Applied Science and Technology*. April 2014. 4 (5): 121–135.

Satter, Abdus, James E. Varnon, and Muu T. Hoang. "Reservoir Management: Technical Perspective." SPE. Paper 22350, SPE, International Meeting on Petroleum Engineering, Beijing, China, 1992 March. 24–27.

Wenzel, Friedemann. *Perspectives of Modern Seismology*, Springer, Berlin. 2005 January. 169–185.

Wison, Alam. Project Tests High-Performance Cloud Computing for Reservoir Simulations. *Journal of Petroleum Technology*. 2016 July. 68: 7–58.

10 Distributed Digital Control Systems for Oil and Gas Operations

10.1 EVOLUTION OF PROCESS CONTROL OPERATIONS

The field of operations management has evolved significantly over the years. It originated during the Industrial Revolution in the 18th century, when the steam engine automated production processes were introduced. In the early 20th century, scientific management techniques, such as time and motion studies, were employed to optimize workflows. Henry Ford later revolutionized production by implementing assembly lines and interchangeable parts.

As sizeable industrial process plants developed, the control systems also underwent various stages of evolution. Initially, control was managed from dispersed panels within the process plant, which required significant manpower and did not offer an overall view of the process. Later on, when process plant data was transmitted to a central control canter, consolidating the remote panels and reducing the need for more manpower while providing a more comprehensive overview of the processing system, however, this setup had limitations as each control loop had its controller hardware, necessitating operators to move around the control room to monitor different process parts.

The introduction of electronic processors and graphic display equipment allowed the replacement of old discrete controllers with computer-software-based algorithms. These algorithms were hosted on a network of input/output (I/O) racks with control processors located around the plant and communicated with graphic displays in the control room or rooms, creating distributed control systems.

Digital control systems (DCSs) brought about flexible interconnections and reconfigurations of process controls, such as cascaded loops and interlocks. DCSs also enabled fast alarm handling/trending, automatic event logging, reduced need for physical records, networked control racks inside the plant to reduce cabling runs, and provided high-level overviews of plant status and production levels.

The reliability of DCS is due to the distribution of the control processing around nodes in the system. This will remove a single processor failure. If the processor fails, it will only affect one part of the plant processing, as opposed to a loss of a central computer which may affect most processes. The computing power local to the field I/O connection racks also ensures fast controller processing times by removing possible network and central processing delays. Digital control systems (DCSs), which are microprocessor-based, comprise a high-resolution host computer and multiple control loops. Using communication from multi-loop controllers, a communication

DOI: 10.1201/9781003307723-10

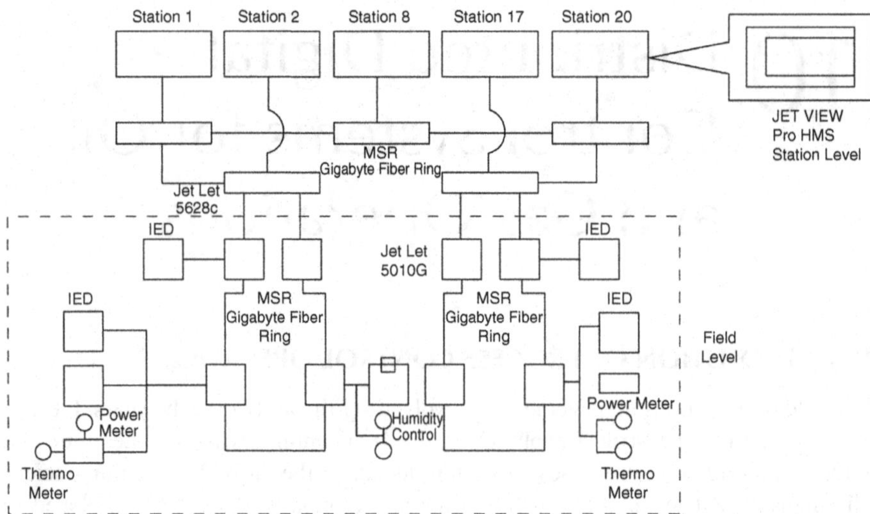

FIGURE 10.1 Manufacturing control of multiple product system diagram.

network is established. It is known as a data highway or node. Functional-level manufacturing plant control logic is explained in the flow chart shown in Figure 10.1.

A distributed control system is a computerized control system for any latest state-of-the-art technology-based industries, including oil and gas process, plant, and drilling rig, usually with many control operational loops. Autonomous controllers are distributed throughout the system. However, there is no central operator supervisory control. This contrasts with systems that use centralized controllers, either discrete controllers located at a central control room or within the main computer. The DCS concept reduces installation operating costs with remote monitoring and supervision by centralizing control functions near the process plant.

The evolution of DCS technology has produced a modern, graphical, alarm trending, highly interactive integration platform, which provides process control functionality and real-time data connectivity between the process plant control room and the operational management. DCS is a new breed of scalable and flexible control system designed to provide end-users in oil and gas and petrochemical process systems, and similar process systems with an automation solution that can fit their applications are suitable than programmable logic controllers (PLCs). Technology advancement is driving innovations, and industrial organizations need to stay in step to remain competitive. This includes effective solutions for automating many plant processes and delivering the information required to make effective operational and business decisions.

10.1.1 Background

Early days, minicomputers were used in the control of industrial processes since the beginning of the 1960s. After the IBM-1800 model, an early computer had I/O hardware to collect process signals in a process plant for conversion from field contact levels and analog signals to the leading digital domain.

The industrial control computer system was built first in 1959 at the Texaco Port Arthur, Texas, refinery with an RW-300 of the Ramo-Wooldridge Company. In 1975, Honeywell and Japanese electrical engineering firm Yokogawa introduced DCS's – TDC 2000 and CENTUM systems. Subsequently, before the PLC-based control system, the single loop programmable control system and usage of solid-state relay-based control technology prevailed for a prolonged time in the industries, particularly for a different type of equipment control with monitoring of analog data locally.

The DCS was introduced mainly due to the increased availability of microcomputers and microprocessors in the process control system. Computers process automation for some time in the form of both direct digital control (DDC) and set point control. Taylor Instrument Company developed the 1010 system, Foxboro the FOX1 system, Fisher Controls the DC^2 system, and Bailey Controls the 1055 systems. This system first appeared in the industry during the 1970s and was used with minicomputers such as the DEC PDP-11, Varian Data Machines, and MODCOMP, which were connected to proprietary I/O hardware. The approach was mainly based on set point control, where process computers supervised clusters of analog process controllers. Workstations typically provided visibility into the process using text and basic character graphics. The availability of a fully functional graphical user interface was limited at that time.

10.1.2 The Chronological Development of Networks

In the 1980s, various users began to explore the possibility of DCSs as more than just primary process control. DDC was designed by the Australian business company Midac in 1981–1982 using R-Tec Australian designed hardware.

As a result, suppliers also start to adopt an Ethernet-based network system with their proprietary protocol layers. Suppose it didn't implement the total TCP/IP standard at that time. Still, the use of Ethernet made it possible to implement the first instances of object management and global data access technology. The first PLCs-based integrated system came in 1980 and was introduced into the DCS infrastructure system. Foxboro is the first DCS supplier to adopt UNIX and Ethernet networking technologies, introducing the I/A Series system in 1987.

10.1.3 Some of the Significant Development of the Network System in the 1990s

The drive toward the beginning of the 1980s gained great momentum through the 1990s. The most significant transition undertaken was moving from the UNIX operating system to the Windows environment. While the realm of the real-time operating system for control applications remains dominated by real-time market variants of UNIX or proprietary operating systems, everything above real-time control has made the transition to Windows.

Microsoft introduced the desktop and server layers resulted in the development of technologies such as OLE for process control through Open Protocol Communication (OPC) system, which became a de facto industry connectivity standard. Internet technology also began to make its footprint in automation and the world, with most DCS-human–machine interface (HMI) Internet connectivity. The "Fieldbus Wars" started where similar organizations competed to become the IEC Fieldbus standard for digital communication with field instrumentation instead of a 4–20-mA analog communications system. Recently, this technology developed significant momentum, with the market consolidated around Ethernet I/P, Foundation Fieldbus, and Profibus PA for process automation applications.

Automatic control typically involves the transmission of signals or commands/information across the different layers of the system and the calculation of control actions resulting from decision-making. The term DCS stands for the distributed digital control system (earlier it was known as DIDC). They use digital encoding and transmission of process information and commands. DCS is applicable not only for all advanced control strategies but also for the low-level control loops. In the beginning, plants used local, significant case pneumatic controllers; these later became miniaturized and centralized onto control panels and consoles. Their appearance changed very little when analog electronic instruments were introduced. The first applications of process control computers resulted in a mix of the traditional analog and the newer DDC equipment located in the same control room. This mix of equipment was not only cumbersome but also relatively inflexible because the changing of control configurations necessitated changes in the routing of control wires.

10.2 BASIC ELEMENTS OF DISTRIBUTED CONTROL SYSTEMS

DCS continuously interacts with the processes in process control applications once it gets instruction from the control operator. It also facilitates variable set points and the operator's opening and closing valves (the final control element) for manual control. Its HMI, faceplates, and trend display effectively monitored industrial processes.

10.2.1 Elements of DCS

10.2.1.1 Engineering PC or Controller

The main components of a DCS include an engineering workstation, which is the entire system's supervisory controller; an operating station/operating interface unit, or HMI, which is a monitoring device that operates, monitors, and controls the plant's parameters; and a field control station/process control unit, which acquires information from field devices like transmitters, sensors, analyzers, etc. This controller is the supervisory controller over all the distributed processing controllers. Control algorithms and configuration of various devices are executed in this controller. A simplex or redundant configuration could implement network communication between processing and engineering PC.

10.2.2 Distributed Controller or Local Control Unit

Using distributed controller, near-field devices (sensing elements and final control elements) or specific locations where these field devices are connected via the communication link can be located. Distributed controller receives the instructions from the engineering station like set point and other parameters and directly controls field devices.

This system can sense and control analog and digital inputs/outputs by analog and digital I/O modules. These modules are extendable according to the number of inputs and outputs. Distributed controller collects the information from discrete field devices and sends this information to operating and engineering stations.

The controllers act as a communication interface between field devices and engineering stations. In most cases, these act as local control for field instruments.

10.2.3 Operating Station

It is used to monitor entire plant operating parameters and log the data in plant database systems. The trend displays various process parameters and performance of multiple controllers and machines, which provides effective collection and easy monitoring of the operation process. These operating stations are of different types, such as some operating stations (PCs) used to monitor only parameters, some for only trend displays, some for data logging, and some for alarming requirements. These can also be configured to have control capabilities.

10.2.4 Communication Media and Protocol

Communication media consists of transmission cables to transmit the data, such as coaxial cables, copper wires, fiber optic cables, and sometimes wireless. Communication protocols selected depend on the number of devices to be connected to this network.

One example is that RS-232 serial port supports only two devices, and Profibus supports 126 devices or nodes. Some of these protocols include Ethernet, DeviceNet, the Foundation Fieldbus, Modbus, and CAN.

In DCS, two or more communication protocols are used between two or more areas, such as field control devices and distributed controllers, and between distributed controllers and supervisory control stations such as operating and engineering stations.

10.3 ARCHITECTURE OF DCS

DCS is reliable due to the distribution of the control processing around nodes in the system. If one processor fails, it will only affect one section of the plant process operation, as opposed to a loss of a central computer, which would affect the whole process. This distribution of computing local power to the field I/O connection racks (Marshaling racks) also ensures fast controller processing times by removing possible network and central processing delays.

The diagram in Figure 10.1 is a general typical DCS control model that shows functional manufacturing levels using a computerized control system in the manufacturing process. This network diagram shows the controlling process from the central computer display station to the field level station (field instruments).

10.4 IMPORTANT FEATURES OF DIGITAL CONTROL SYSTEMS

In process automation, PLCs control and monitor process parameters at high speed to handle complex processes. However, due to the limitations of several I/O devices, PLCs cannot handle very complex structures.

10.4.1 Handling Complex Processes

Based on the added advantages and reliability, DCS is preferred for complex control applications with more I/Os with dedicated controllers. These are used in manufacturing processes where multiple products are designed using various procedures such as batch process control.

10.4.2 Digital Control System Redundancy

DCS facilitates system availability when needed by redundant features at every level. Resuming the steady-state operation after any outages, whether planned or unplanned, is somewhat better compared to other automation control devices. Redundancy raises the system reliability by maintaining system operation continuously even in some abnormalities while the system is in process. The DCS model was the inclusion of control function blocks as central control. Function blocks evolved from early, more primitive DDC concepts of "Table Driven" easy software. One of the first embodiments of object-oriented software, function blocks were self-contained "blocks" of code that emulated analog hardware control components and performed tasks that were essential to process control, such as the execution of proportional integral derivative (PID) algorithms. Function blocks continue to endure as the predominant control method for DCS suppliers and are supported by key technologies such as Foundation Fieldbus.

Digital communication between distributed controllers, workstations, and other computing elements (peer-to-peer access) was one of the primary advantages of the DCS. Attention was on the networks, which provided the all-important lines of communication that, for process applications, had to incorporate specific functions such as determinism and redundancy.

These distributed controllers are interlinked to both field devices and operating Port of Communication System (PCS) through high-speed communication networks, as explained in Figure 10.1.

Discrete field devices such as sensors and actuators are directly connected to input and output controller modules through a communication bus. These field devices or intelligent instruments can communicate with PLCs or other controllers while interacting with real-world parameters like temperature, pressure, etc.

10.4.3 DCS Architecture

These controllers are distributed geographically in various sections of the control area. They are connected to operating and engineering stations used for monitoring, logging, alarming, and controlling via another high-performance computing (HPC) bus. More communication protocols can be identified as HART, Profibus, the Foundation Fieldbus, Modbus, etc. DCS displays information to multiple displays for the user interface. In DCS, two or more communication protocols are used, such as between field control devices and distributed controllers and between distributed controllers and supervisory control stations such as operating and engineering stations. Some of the critical features of the distributed control system are listed below.

10.4.4 To Handle Complex Processes

In the factory automation system, PLC controls and monitors the process parameters at high-speed requirements. This had a limitation in the number of I/O devices; PLCs cannot handle complex structures. DCS offers many algorithms, more standard application libraries, and pre-tested and pre-defined functions to deal with large complex systems. However, programming to control various applications is easy and consumes less time to program and maintain.

Powerful programming languages: It provides more programming languages like a ladder, function block, sequential, etc., for creating custom programming based on user interest.

10.4.5 More Sophisticated Human–Machine Interface

Similar to the Supervisory Control and Data Acquisition (SCADA) system, DCS can also monitor and control through HMIs, which provides sufficient data to the operator to charge over various processes. It acts as the heart of the system. But this type of industrial control system covers large geographical areas, whereas DCS covers the confined spaces. DCS completely takes the entire process plant to the control room as a PC window. Trending, logging, and graphical representation of the HMIs give an effective user interface. The powerful, alarming system of DCS helps operators to respond more quickly to the plant conditions.

10.4.6 Scalable Platform

The structure of DCS is scalable based on the number of I/Os from small to large server systems by adding more clients and servers in the communication system and adding more I/O modules in distributed controllers.

10.4.7 System Security

Access to control the various processes leads to plant safety. DCS design offers a perfect secured system to control system functions for better factory automation control. Security is provided at different levels such as engineer level, entrepreneur level, and operator level.

10.4.8 Application of Distributed Control Systems

Simple application like load management can be developed in DCS using a network of microcontrollers. The input is supplied from a keypad to a microcontroller, which communicates with the other two microcontrollers. One of the microcontrollers displays the status of the process and the loads, while the other microcontroller always tries to control the relay driver. The relay driver, in turn, drives the relay to operate the load.

10.5 MODERN DCS SYSTEMS

The following are the three new technologies in modern DCS:

- Wireless systems and protocols
- Remote transmission, logging, and data historian
- Mobile interfaces and controls

10.5.1 Embedded Web-Servers

Increasingly, and ironically, DCS are becoming centralized at process plant-level operation, with the ability to log into the remote equipment. It enables an operator to control the enterprise level (macro) and the equipment level (micro). The importance of the physical location drops due to interconnectivity primarily thanks to wireless and remote access.

The more wireless protocols are developed and refined, the more they are included in DCS. DCS controllers are often equipped with embedded servers and provide on-the-go web access. Whether DCS will lead Industrial Internet of Things (IIoT) or borrow critical elements is discussed in the earlier chapter.

Modern DCSs also support neural networks and fuzzy logic applications. The evolution of the existing system to a new system state for optimally distributed controllers, which optimizes a certain H-infinity or the H2 control criterion, is explained in the flow chart in Figure 10.2.

In a pre-DCS era central control room, while the controls are centralized in one place, they are still discrete and not integrated into one system.

In DCS control room, plant information and rules are displayed on computer graphics screens. The operators are seated and can monitor and control any part of the process from their screens while retaining a plant overview.

10.5.2 Five Subsystems of Distributed Control Systems

Process interface is responsible for collecting data from measurement instruments and issuing signals to actuating devices such as pumps, motors, and valves.

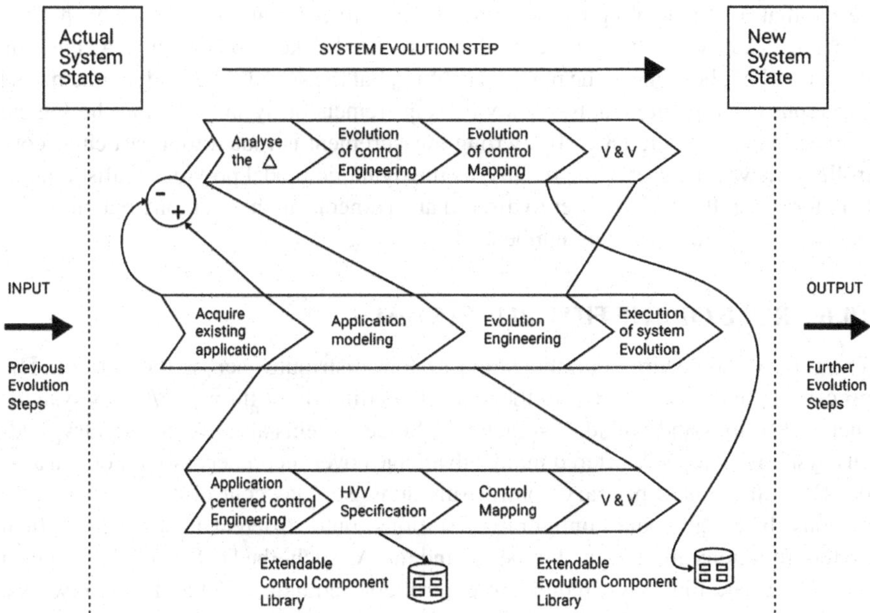

FIGURE 10.2 System evolution step from actual present system to new system.

Process control is responsible for translating the information collected from the process interface subsystem and determining the signals sent to the process interface subsystem based on pre-programmed algorithms and rules set in its memory.

Process operations are responsible for communicating with operations personnel at all levels, including operator displays, alarms, process variables, activities trends, summary reports, and operational instructions and guidelines. It also tracks process operations and product batch lots.

Application engines are the repository for all of the programs and packages for the system from control, display and report configuration tools to program language compilers and program libraries to specialized containers such as database managers, spreadsheets, and optimization or expert system packages repositories for archived process information.

Communications subsystems enable information flow between various DCS subsystems and other computerized systems such as laboratory information management systems, plant inventory management, plan scheduling systems such as MRP II, and plant maintenance systems.

Integrating these systems into a cohesive whole has dramatically increased the level of automation possible to improve quality and productivity.

10.5.3 Distributed Control System Checkout

Distributed control system (DCS) checkout alone will not warrant the construction of a dynamic model of a plant. But if a model is available, the modifications needed to run a DCS checkout are relatively small. The purpose of the DCS checkout is to verify that all the cabling connecting the DCS to the plant and the DCS internal Technical Advisory Group (TAG) allocations are hooked up correctly. A dynamic model will not help check the physical cabling, but the signals from a dynamic model can replace the plant signals. This will help tremendously in verifying the logical connections inside the DCS. If a wrong measurement is routed to a particular controller, this will be seen quite readily, as the dynamic model provides realistic numbers for these. It is much easier to discern an erroneous number among real numbers than to match quasi-random numbers.

10.6 ROLE OF THE FIELDBUS SYSTEM

The robust and secure communication system distributes across the control. The process inputs are connected to the controllers directly or through I/O bus systems such as Profibus and Foundation Fieldbus. Some systems also use proprietary fieldbus systems. The DCS offered many advantages over its predecessors. For starters, the DCS distributed primary control functions, such as controllers, I/O, operator stations, historians, and configuration stations, onto different boxes. The critical system functions are designed to be redundant. As such, the DCS tended to support redundant data highways, redundant controllers, redundant I/O and I/O networks, and, in some cases, redundant fault-tolerant workstations. In such configurations, if any part of the DCS fails, the plant can continue to operate. The evolution of communication technology and the supporting components has dramatically altered the fundamental structure of the control system. Communication technology such as Ethernet and Transmission Control Protocol (TCP)/User Data Protocol (UDP)/Internet Protocol (IP) combined with standards such as OPC allowed third-party applications to integrate into the control system. Also, the general acceptance of the object-oriented design, software component design, and supporting tools for implementation has facilitated the development of better user interfaces and the implementation of reusable software.

With advancing technologies, DCS has rapidly expanded its features, functions, performance, and size. The DCSs available today can perform very advanced control functions, along with the powerful recording, totalizing, mathematical calculations, and decision-making functions. The DCS can also tailor to carry out particular functions, which the user can design. An essential feature of modern-day DCS is integrating enterprise resource planning (ERP) and IT systems through exchanging various pieces of information.

It is desirable to review the evolution of control systems to understand the DCS. This includes hardware elements, system implementation philosophies, and the drivers behind this evolution. This will help understand how process control, information flow, and decision-making have evolved over the years.

In analog controls systems, instruments produce a 4–20 mA output signal that travels from the remote field areas to the control room through marshaling rack, hidden I/O cards, or Remote Telemetry Unit (RTU) over twisted pair of cables.

Similarly, 4–20 mA control signals travel from the control system to valve actuators, pumps, or other final control devices. Hundreds, sometimes thousands, of cables are laid through cable trays, termination racks, cabinets, enclosures, and conduits (Figures 10.3 and 10.4).

FIGURE 10.3 Terminations.

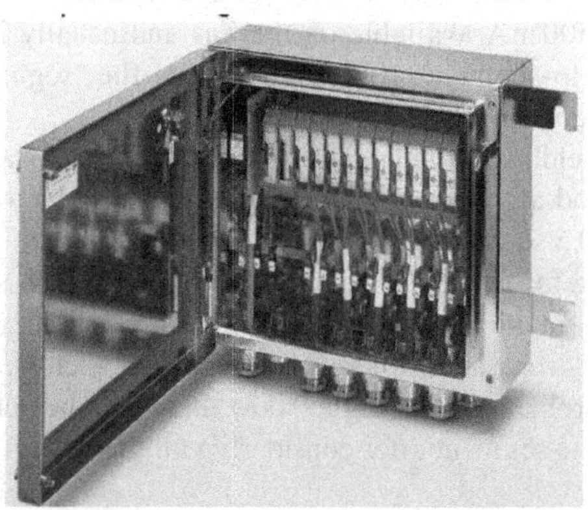

FIGURE 10.4 Traditional 4–20-mill ampere field-wiring fieldbus installation that simply often results in a rat's nest of wire cables.

10.6.1 Foundation Fieldbus

The availability of a Foundation Fieldbus, which is a potent processor suitable for field instrumentation, has provided the way to remove the bulk of these cables and, at the same time, enhance data available from the plant. Instead of running individual lines, Fieldbus allows multiple instruments to use a single thread, called a "trunk" or a "segment" (Figure 10.4); each device connects to the line as a "drop." Instruments, of course, must have a Fieldbus interface to connect to the segment and some type of software running to provide the Fieldbus communications.

A Fieldbus trunk or segment—either Foundation Fieldbus H1 or Profibus—is a single twisted-pair wire carrying both a digital signal and DC power that connects up to 32 Fieldbus devices (temperature, flow, level, and pressure transmitters, smart valve positioners, actuators, etc.) to a DCS or similar control system. Most devices are two-wire bus-powered units requiring 10–20 mA, but it is also possible to have 4-wire Fieldbus devices, typically where a device has an exceptionally high current draw.

10.6.2 Case Study-1

The Fieldbus segment begins at an interface device at the control system. On a Foundation Fieldbus H1 (Friendly Fire (FF)) system, the interface is called an H1 card; on a Profibus Public Address system (PA), it is a profit bus DP/PA segment coupler. In terms of signal wiring and power requirements for the segment, FF and PA are identical:

- Minimum device operating voltage of 9 V
- Maximum bus voltage of 32 V
- Full cable length of 1,900 m (shielded twisted pair)

The DC power required by the bus usually is sourced through a Fieldbus power supply or "power conditioner," which prevents the high-frequency communication signal from being shorted out by the DC voltage regulators. Typical power conditioners make 350–500 mA available on the bus and usually incorporate isolation to prevent segment-to-segment cross-talk. For PA, the "segment coupler" usually includes the power conditioning component.

In Foundation Fieldbus segments, the power conditioners are separate from the H1 interface card and are often installed in redundant pairs to improve the overall reliability. Figure 10.5 shows a typical Fieldbus segment.

When calculating for no devices that can fit on a Fieldbus segment, a user must consider the maximum current requirement of each machine, the length of the part (because of voltage drops along the cable), and other factors. The calculation is a simple Ohm's law problem, intending to show that it can be delivered at least 9 V at the farthest end of the segment after considering all the voltage drops from the total segment current.

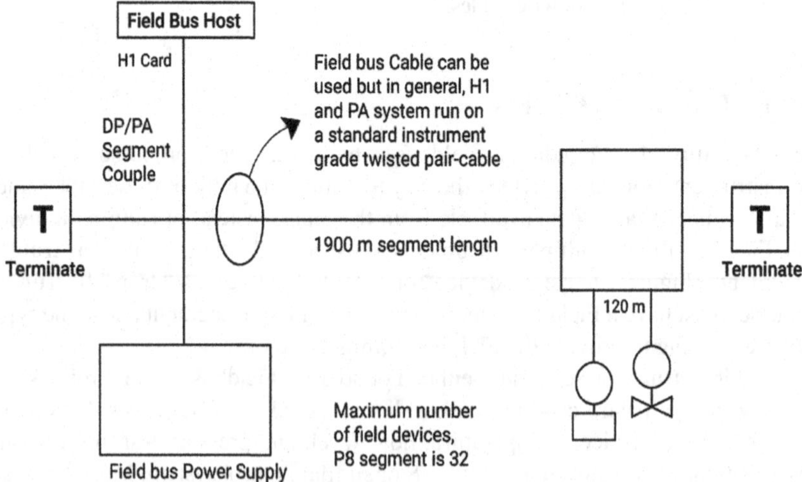

FIGURE 10.5 Foundation Fieldbus network.

10.6.3 Case Study-2

For example, driving 16 devices at 20mA each requires 320mA, so if the segment is based on an 18AWG cable (50 Ohms/km/loop) with a 25V power conditioner, the maximum cable length is 1,000m to guarantee 9V at the end.

Note that many users also specify a safety margin on top of the 9V minimum operating voltage to allow for unexpected current loads and adding additional devices in the future.

10.6.4 Connecting Instruments

As noted, each Fieldbus device connects to the segment in parallel via a "drop" on the Fieldbus segment called a spur. The most direct spur connection is a "T." The problem with simple "T" connections is that if any of the devices or cables are shorted out, it takes down the entire segment (Figure 10.6).

A short can occur during field maintenance of an instrument, from an accident, corrosion causing electrical problems, or a host of other possibilities. Short-circuit protection is, therefore, a requirement for proper Fieldbus implementation.

Another way to connect Fieldbus devices is via junction boxes designed explicitly for Fieldbus, often referred to as "device couplers."

When a short circuit deprives other instruments of power, some may "drop off" the segment because they do not have enough ability to operate correctly.

Consequently, when current limiting protection is used in a device coupler, many end-users allow a safety margin. They do not install as many instruments as the segment can theoretically power; instead, they leave a certain number of spurs empty.

For example, if a user wants the segment to keep working with two failures—which can draw up to 120mA of current—the segment calculations must assume a maximum current availability of 350mA minus 120mA for the faults value of

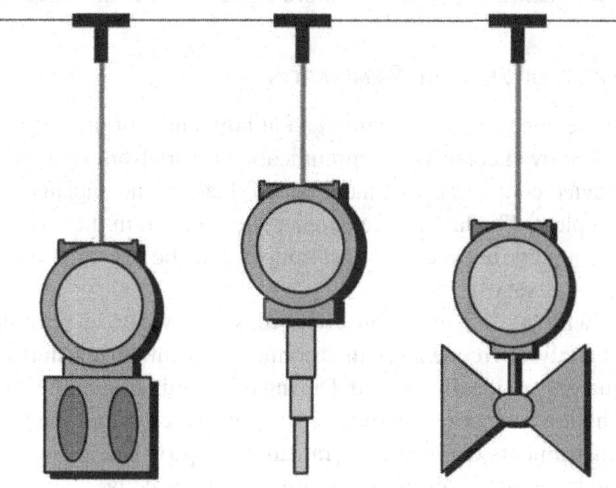

FIGURE 10.6 "T" configurations are the most straightforward Fieldbus connection. However, if one device fails or short circuits, it "takes down" the entire segment.

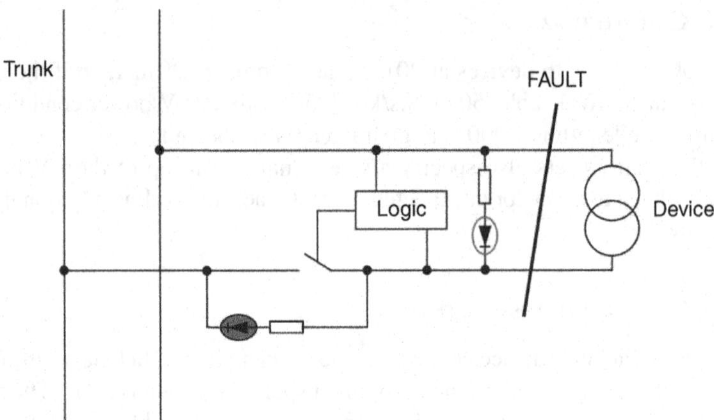

FIGURE 10.7 Fold-back short-circuit protection coupler line diagram.

230 mA. Instead of theoretically powering 32 devices that draw 10 mA each, the segment can only support 23 such devices. In practice, some users are wary of relying on current limiting couplers, and most couplers limit each component to only 16 machines to prevent large-scale segment failures.

Fold-back short-circuit protection has logic that detects a short, removes the shorted circuit from the segment, and lights an LED. This prevents a short from affecting the segment explained in Figure 10.7. As used in device couplers, the fold-back technique disconnects the shorted spur from the part, thus preventing the loss of an entire segment. The fold-back method has a logic circuit on each motivation (Figure 10.6) that detects a short in an instrument or spur, disconnects that spur from the segment, and illuminates a red LED that can be seen by maintenance personnel.

With fold-back device couplers, users no longer have to worry about spur failures and can have confidence about placing more devices on Fieldbus segments.

10.6.5 Process of Segment Termination

Every Fieldbus segment must be terminated at both ends for proper communication. If a part is not removed correctly, communication errors from signal reflections may occur. Most device couplers use manual on/off dual in-line package (DIP) switches to complete couplers. The last device coupler should contain the terminator in a segment, and all couplers between the last couple and the H1 card should have their terminator switches set off.

The boxes with a "T" in Figure 10.6 illustrate where a specific segment is terminated correctly. A frequent commissioning problem during startup is determining that terminators are ideally located. During the installation of the Fieldbus system, the DIP termination switches sometimes are set incorrectly, creating problems during startup. The instruments can behave erratically, drop off the segment mysteriously, and raise havoc because the terminations are not set correctly.

Sometimes, diagnosing the problem requires examining each device coupler to determine if the switches are correctly set throughout the segment.

Automatic segment termination, as found in FF device couplers, simplifies commissioning and startup. It automatically activates when the device coupler determines that it is the last Fieldbus device coupler in the segment; if it is, it terminates the segment correctly. It removes the part if it is not the final device since the downstream device coupler will assume that responsibility. No action—such as setting DIP switches—is necessary by the installation person to terminate a segment properly.

If a device coupler is disconnected from the segment accidentally or for maintenance, the automatic segment termination detects the change and terminates the part at the proper device coupler. This allows the remaining devices on the segment to continue operation.

10.6.6 Advantages of Fieldbus Communication

The considerable cost savings initially justified Fieldbus due to the reduction in huge cable requirements and maintenance costs. Fieldbus often required only a few dozen segments or trunks instead of running hundreds or thousands of wires.

The savings from running cables can be offset by the high cost of Fieldbus components and the reluctance of many users to install all the instruments possible on a segment. Being forced to provide for short circuits, for example, limits the number of devices that can be put on a part. In many cases, it's a toss-up from a hardware cost and labor perspective. In other cases, hardware cost savings become more realistic when one can use them in full capability.

The true advantage of Fieldbus is its ability to diagnose equipment problems, cut maintenance costs, provide information for asset management, allow control at the device level, and allow the use of intelligent devices.

One oil company in Alaska did a study of savings, and it was done by using a Foundation Fieldbus system. The savings included:

Wiring: They could have achieved a 98% reduction in the home-run wiring because, with Fieldbus, they could eliminate the costly maze of wiring between each remote field instrument and the control room. Also, they reduce terminations by 84%.

Control room: Fewer terminations also freed up two-thirds of the cabinet space that would require traditional technology.

10.6.6.1 Commissioning of System

Field checkout and QA/AC time was reduced by 83%. Installation of each transmitter took only 20 minutes rather than the 2 hours needed with non-fieldbus technology.

10.6.6.2 Engineering Drawings

The effort required for new drawings when adding oil wells has been reduced by 92% because of the Foundation Fieldbus and the host system's configuration tools and object-oriented capabilities.

The oil and gas industry was the first to embrace Fieldbus fully and now installs it at many new refineries, offshore platforms, and other facilities.

10.7 COMPARISON BETWEEN FOUNDATION FIELDBUS H1 VS. PROFIBUS PA

From a field-wiring perspective, Foundation Fieldbus and Profibus are physically identical. They use the same twisted-pair cables and device couplers and require the same segment terminators terminology. Both handle up to 32 devices per segment.

One primary difference is that Profibus is a polling system, while Foundation Fieldbus Fieldbus utilizes cyclic transmission. Other differences include the following:

- FF devices have scheduled times to transmit their information, whereas PA devices submit their data at random times. In PA, slave polling involves the bus master asking for information from the devices. The active link scheduler in FF has a timetable, determining when devices communicate on the segment.
- Address allocation in PA must be done by communicating with each device individually, whereas FF devices will announce themselves to the bus master. Each PA device will return to the bus master with its information and transmit the relevant data to other machines. FF devices can talk to each other and bypass the bus master, hence providing peer-to-peer communication.
- FF devices can have built-in function blocks that allow them to talk to each other peer-to-peer, perform control functions, and continue to operate if communications are lost to the control system. Profibus systems do not have function block capability. Profibus instrumentation reports to and takes directions from the PA master; if communications to the PA Master(s) are lost, the instruments must go to a fail-safe position or maintain their last settings until directed otherwise.
- FF and PA differ in how the segment control cards connect to the DCS or control system. FF uses a high-speed Ethernet network to connect remote H1 cards to the DCS; PA uses Profibus DP, an RS-485 port protocol network, or PROFINET, an Ethernet-based network, to join its PA devices to the DCS.

10.8 STRUCTURE OF DCS

A typical DCS control line diagram is depicted in Figure 10.8.

Various levels of DCS control system are categorized below for a better understanding:

- **Level 0** contains the field devices such as flow and temperature sensors and final control elements, such as control valves.
- **Level 1** contains the industrialized I/O modules, and they are associated with distributed electronic processors.

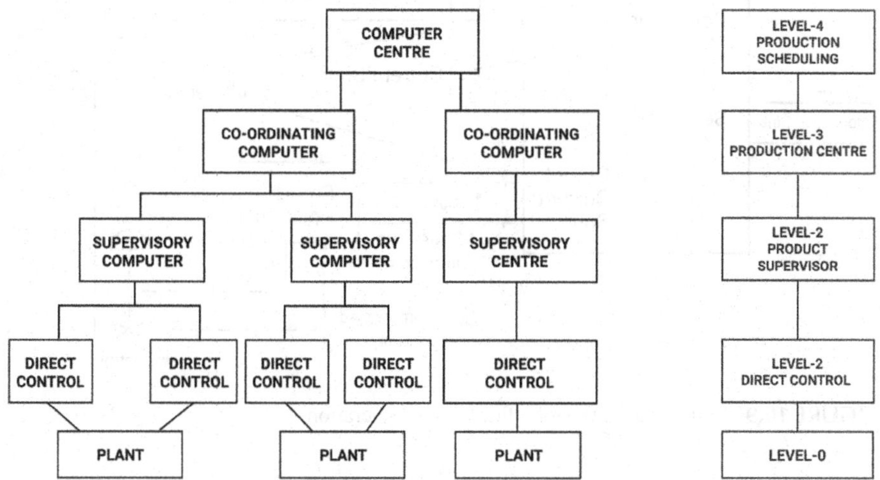

FIGURE 10.8 DCS control line diagram.

- **Level 2** contains the supervisory computers, which collect information from processor nodes on the system and provide the operator control screens.
- **Level 3** is the production control level, which does not directly control the process but is concerned with monitoring production and monitoring targets.
- **Level 4** is the production scheduling level.

Levels 1 and 2 are the functional levels of a traditional DCS, in which all equipment is part of an integrated system from a single manufacturer.

Levels 3 and 4 do not strictly process control in the traditional sense, but production control and scheduling occur.

The diagram shown in Figure 10.9 is a general model that shows functional levels using computerized flow control logic.

The schematic drawn above explained that Flow Controller shall send signal to flow control valve via a smart positioner.

10.8.1 Functional Input

The processor nodes and operator graphical displays are connected over proprietary or industry-standard networks, increasing network reliability by dual redundancy cabling over diverse routes. This distributed topology also reduces the amount of field cabling by siting the I/O modules and their associated processors close to the process plant—an example of a continuous flow control loop. Signaling is by industry-standard 4–20 mA current loops, and a "smart" valve positioner ensures that the control valve operates correctly.

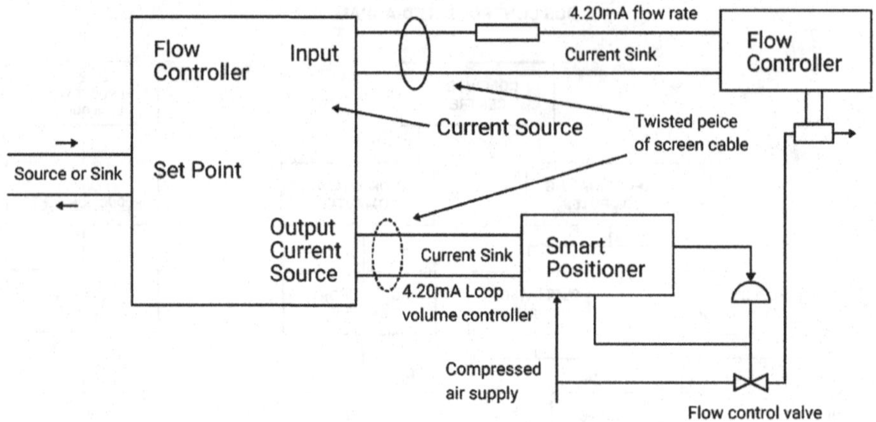

FIGURE 10.9 Functional levels of a flow control operation.

The processors receive information from input modules, process the data, and decide control actions signaled by the output modules. The field inputs and outputs can be analog signals, e.g., 4–20 mA DC loop or two-state signals that switch either "on" or "off," such as relay contacts or semiconductor switches.

DCSs are connected to sensors and actuators and use set point control to control material flow through the plant. A typical application is a PID controller fed by a flow meter and using a control valve as the final control element. The DCS sends the set point required by the process to the controller, which instructs a valve to operate so that the process reaches and stays at the desired set point (see 4–20 mA schematic, for example).

Large-size oil refineries and central oil and gas processing plants have several thousand I/O points and massive DCS. Processes are not limited to fluidic flow through pipes, however, and can also include things like paper machines and their associated quality controls, variable speed drives and motor control centers, cement kilns, mining operations, ore processing facilities, and many others.

DCSs in very high-reliability applications can have dual redundant processors with "hot" switch over on fault to enhance the reliability of the control system.

Although 4–20 mA has been the primary field signaling standard, modern DCS systems can also support Fieldbus digital protocols, such as Foundation Fieldbus, Profibus, HART, Modbus, and PC Link, and other digital communication protocols such as Modbus.

10.8.2 Supervisory Control and Data Acquisition

A distributed control system (DCS) is used to control production systems within the exact geographic location. It generally involves a computer that communicates with control elements distributed throughout the plant or process, e.g., machine or process controllers and PLCs, through a bus or directly and displays gathered data. DCS systems have a hierarchy of controllers distributed through a plant and connected by a communications network for command and monitoring, whereas SCADAs have central control. SCADA systems often contain DCS components. An example of a DCS system could be an oil refinery where there are many control systems such as flow controllers and closed-loop controllers by which valves are operated to obtain set values.

A system that allows an operator, in a location central to a widely distributed process, such as oil or gas field, pipeline system, to make set point changes on distance process controllers, open or close valves or switches to monitor alarm, and to gather measurement information.

SCADA system is similar to a distributed control system except for sub-control systems being geographically dispersed over a large area and accessed using remote terminal servers. A distributed control system (DCS) is a supervisory control system that typically controls and monitors set points to sub-controllers distributed geographically throughout a factory.

SCADA systems are used for controlling, monitoring, and analyzing industrial devices and processes. This network, which is the platform system, provides the capability to measure and control specific elements of the first system.

The nature of SCADA has led to different views as to whether it forms part of the IIoT ecosystem. For example, discussion of SCADA system forensic analysis within IIoT contrasts with a statement that SCADA is simply the predecessor to IIoT significantly as SCADA systems had evolved to contact the internet but do not have the analytics and level of connectivity that are available in IIoT.

The evolution and commoditization of Distributed Control System technology have produced a modern, graphical, highly interactive integration platform, which provides process control functionality and real-time data connectivity between the plant floor and the enterprise. Today, DCS is a new breed of scalable and flexible control system designed to provide end-users in oil and gas and petrochemical processes and similar process industries with an automation solution that can fit their applications better than PLCs and costs less than a traditional DCS. Technology advancement is driving innovations, and industrial organizations need to stay in step to remain competitive. This includes effective solutions for automating many plant processes and delivering the information required to make effective operational and business decisions.

New electronic versions gradually superseded DCS systems. Further development resulted in modern control systems that utilize digital Fieldbus networks (HART, FF, etc.) to gather much more data from field devices than a single temperature parameter or provide a set point. More recently, the availability of Ethernet field instrument devices has allowed for a more standard network throughout the system architecture. One of the most significant milestones in DCS technology is building a control system with a required network structure that provides information into a single platform. Connecting this to the overall enterprise offers access to much more significant and more complex information.

10.9 ESD/DCS INTERFACES

Distributed control systems (DCSs) and Emergency Shut Down (ESD) are designed; they are functionally segregated such that a failure of the DCS does not prevent the ESD from shutting down and isolating the facilities. Alternatively, loss of the ESD system should not prevent an operator from using the DCS to shut down and isolate the system. There should be no executable commands over the ESD–DCS communication links. Communication link should only be used for bypasses, status information, and the transmission of reports. The confirmation of ESD reset action can be incorporated into the DCS, but actual reset capability should not.

10.10 REAL-TIME OPTIMIZATION (RTO) TECHNOLOGY

RTO is a category of closed-loop process control that optimizes process performance in real time for systems. Compared to traditional process controllers, they are different as they are generally built upon model-based optimization systems and are usually large scale. RTO refers to techniques allowing the continuous evaluation and manipulation of process operating conditions to maximize economic productivity. RTO tries to predict properties, such as a product's characteristics, by utilizing the plant operating conditions. This is done by first selecting a process and then formulating a suitable problem statement. Optimization of set points involves a total of two models: Economic and Operating.

The economic model comprises an objective function that needs to be minimized or maximized, while the operating model is a steady-state process model containing all variable constraints.

The steps that lead to solving an RTO problem include the following:

- Identifying the process variables
- Selecting the objective function
- Developing the process model and constraints
- Simplifying the objective function and process model
- Computing the optimal solution
- Performing sensitivity studies

An advanced control strategy project can be executed within four stages, given as follows.

Estimation of Benefits: The decision to implement such a strategy shouldn't be without preliminary cost–benefit analysis. This step is vital to ensure that the plant's financials would be able to sustain the process.

Implementation of Algorithms and Process Modeling: The configuration of the real-time database and controller is a must before executing the advanced control algorithms. All the process must be thoroughly tested, and each one must have redundant communication links to prevent failover.

Commissioning of DCS: This process must be initiated once the control system has been interfaced and merged within the existing infrastructure. The communication platform must be tested, after which the multivariable controller should be run module in open-loop mode. Next, each control loop should be tuned, and the controller actions tested and verified.

Maintenance: This phase is required to ensure optimum performance of the advanced controls. Algorithms' performance must be compared daily, and if any discrepancies arise, the controller must be re-tuned accordingly.

10.10.1 Real-Time Optimizer of the Typical Cycle

Oil and gas process industries compelled increasingly to operate profitably to compete in a very dynamic and global market. Increasing pressure on product quality specification requirements, decreasing profit margins, and changing market conditions have given more incentive to operating facilities to use real-time or online optimization (RTO).

Process plant measurements are collected online and checked for steady-state operation. If the plant is steady-state, data reconciliation is performed on the measured data, and the plant process model is updated based on reconciled data.

10.10.2 Typical Functions of a Real-Time Optimizer (RTO)

Rigorous optimization is then carried out using the updated model, economic data, and product requirements to find the new set targets for the plant operating variables. New set points are now passed to the MVC or directly to the distributed control system for implementation on the plant.

10.10.3 RTO Optimization Finds an Optimum at Each Iteration

Distributed control systems are present in most chemical process industries, and computing power for RTO became inexpensive, allowing its cost to drop. The primary issue in RTO use appears to be developing a process model, which requires considerable effort and expertise. Several types of RTO application systems are available in the market, and the following are the major operating process systems:

- Crude oil refinery and hydrocrackers unit.
- Olefins plants: The process models have more than 30,000 equations and variables; one entire cycle of RTO takes about 20 minutes.
- Power plant optimization, which has a potential 3% saving in fuel consumption.
- After installing the DCS system correctly, the plant operation efficiency increased by 3% in the LNG and oil processing plants.

The typical payback period for RTO applications is less than a year. Shortly, RTO could become an integral part of the design of the new plant to optimize plant operation properly. RTO applications provide capabilities for offline optimization of the entire plant or sections of the plants. The users can also use the models for plant operator training and operational problem-solving.

The RTO application monitors the process key measurements, such as plant feed rates, feed compositions, and streamflow, to determine if the plant is at steady-state to run a parameter case.

If the RTO program detects that the plant is at steady-state, averages of all the plant measurements are calculated and used as inputs to the optimization model. The input data are screened using RTO validity-checking programs to set up the specifications for the parameter case. The RTO parameter case then runs to match the model with the current plant operating conditions.

Upon completing the parameter case, the optimization validity program is activated to process the on/off control optimization flags and the current operational constraints. Then, an optimal set of operating variables is calculated by the optimizer. The optimization case calculates the values of operating variables which will maximize the profit objective function.

Before sending the new set of optimal operation variables to the plant DCS, the RTO program checks the status of the plant. The optimization targets are copied to the MVC controllers, which will ensure the actual targets are implemented at speeds specified by the MVC tuning parameters. In general, the targets are ramped as controlled variable limits.

Once these optimal targets are implemented, the RTO system steady-state checker resumes looking at the critical variables in the plant to determine if the plant is steady, and a new RTO cycle starts.

For chemical and petrochemical operations, closed-loop online optimization allows accurate determination of the most profitable key plant variables, subject to operating constraints such as feedstock availability, temperature and pressure limits, and equipment limitations. The optimization results are implemented via advanced control with no human intervention.

10.11 ELEMENTS OF DEPENDABILITY

The following dependability attributes are qualities of a system, and these attributes can be assessed to determine their overall dependability using qualitative measures:

- **Availability**—readiness for correct service
- **Reliability**—continuity of accurate service
- **Safety**—the absence of catastrophic consequences on the user(s) and the environment
- **Integrity**—lack of improper system alteration
- **Maintainability**—the ability for easy maintenance (repair)

As these definitions suggested, only Availability and Reliability are quantifiable by direct measurements while others are subjective. Safety cannot be measured directly via metrics but is a personal assessment that requires critical information to be applied to give a level of confidence, while Reliability can be calculated as failures over time.

Confidentiality, i.e., "the absence of unauthorized disclosure of information" is also used when addressing security. Security is a composite of Confidentiality, Integrity, and Availability. Security is sometimes classed as an attribute, but the current view is to aggregate it with Dependability and treat Dependability as a composite term called Dependability and Security. Practically, applying security measures to the appliances of a system generally improves the Dependability by limiting the number of externally originated errors.

10.12 THREATS PERCEPTION

Threats are things that can affect a system and cause a drop in Dependability. There are three main terms that must be clearly understood:

Fault: A fault (usually referred to as a bug for historical reasons) is a defect in a system. The presence of a flaw in a system may or may not lead to Failure. For instance, although a system may contain a fault, its input and state conditions may never cause this fault to be executed so that an error occurs; and thus, that particular defect never exhibits as a failure.

Error: An error is a discrepancy between the intended behavior of a system and its actual behavior inside the system boundary. Errors occur runtime when some part of the system enters an unexpected state due to the activation of a fault. Since mistakes are generated from invalid conditions, they are hard to observe without unique mechanisms, such as debuggers or debug output to logs.

Failure: A failure is an instance in time when a system displays behavior that is contrary to its specification. An error may not necessarily cause a loss; for example, an exception may be thrown by a plan, but this may be caught and handled using fault tolerance techniques so the overall operation of the system will conform to the specification.

It is important to note that Failures are recorded at the system boundary. They are Errors that have propagated to the system boundary and have become observable. Faults, Errors, and Failures operate according to a mechanism. This mechanism is sometimes known as a **Fault-Error-Failure** chain. As a general rule, when activated, a spot can lead to an error (an invalid state), and the weak state generated by mistake may lead to another error or a failure (an observable deviation from the specified behavior at the system boundary).

Once a fault is activated, an error is created. An error may act in the same way as a fault in that it can produce different error conditions. Therefore, an error may propagate multiple times within a system boundary without causing an observable failure. If a mistake reproduces outside the system boundary, a loss is said to occur. A failure is a point at which it can say that a service is failing to meet its specification. Since it feeds the output data from one service into another, a failure in one service may propagate into another service as a fault to form a chain of the form: Fault leading to Error leading to Failure leading to Error, etc.

10.12.1 Measure of Detection

Since the mechanism of a Fault-Error-Chain is understood, it is possible to construct means to break these chains and thereby increase the Dependability of a system. Four standards have been identified so far:

- Prevention
- Removal
- Forecasting
- Tolerance

1. Fault Prevention deals with preventing faults from being incorporated into a system. This could be accomplished by the use of development methodologies and good implementation techniques.
2. Fault Removal can be sub-divided into two sub-categories: Removal during development and Removal during use. Removal during development requires verification to detect and remove faults before a system is put into production. Once plans have been put into production, a method is needed to record failures and remove them via a maintenance cycle.
3. Fault Forecasting predicts likely faults to be removed or can circumvent their effect.
4. Fault Tolerance deals with putting mechanisms in place that will allow a system to still deliver the required service in the presence of faults, although that service may be at a degraded level.

Dependability means are intended to reduce the number of failures presented to the user of a system. Losses are recorded generally over time, and it is helpful to understand how their frequency is measured to assess the effectiveness of means.

10.13 DEPENDABILITY OF A DCS

Dependability always benefits structured information systems, e.g., with service-oriented architecture (SOA), to introduce a more efficient ability, the survivability, thus taking into account the degraded services that an information system sustains or resumes after a non-maskable failure.

The flexibility of current frameworks encourages system architects to enable reconfiguration mechanisms that refocus the available, safe resources to support the most critical services rather than over-provisioning to build a failure-proof system. The generalization of networked information systems introduced accessibility to give greater importance to users' experience. The primary mechanism used by a fault tolerant unit is the replication of software, hardware, information, and time.

10.14 CONCLUSION

Based on the technology upgradation and advantages of the latest control system with OPC-enabled communication, management information system is also of prime importance toward the decision-making through remote monitoring of parameters/data.

Also, along with the development toward modern state-of-the-art technology-based control systems, safety index limit (SIL) has got its importance in deciding the safety factor (SIL-2/SIL-3/SIL-4) based on the type of industries and its precious control system requirement.

BIBLIOGRAPHY

Boyes, Hugh, Bilal Hallaq, Joe Cunningham, and Tim Watson. The Industrial Internet of Things (IIoT): An Analysis Framework, *Computers in Industry*. 2018. 101: 1–12.

Mehta, B. R. and Y. Jaganmohan Reddy. *Industrial Process Automation Systems Design and Implementation*, Elsevier, Oxford, UK. 2015. 75–133, 401–416.

Rondeau, Patrick, and Lewis A. Litteral. Evolution of Manufacturing Planning and Control Systems: From Reorder Point to Enterprise Resource Planning, *Production and Inventory Management Journal*. 2001. 42 (2): 1–2.

Rouse, Margaret, "Real-Time Optimization." Techopedia. 2017 March 2. https://www.techopedia.com/definition/182/real-time-optimization-rto.

Singh, Shivam. "Basics of Field Bus Foundation." Slide Share. 2013 December 27. https://www.slideshare.net/ShivamSingh59/basics-of-fieldbus.

Trentesaux, Damien. Distributed Control of the Production Systems. *Engineering Applications of Artificial Intelligence*. 2009 October. 22 (7): 971–978.

Verhappen, Ian, and Augusto Pereira. *Foundation Fieldbus*, Fourth Edition, Kindle Amazon Edition, International Society of Auto, Durham, NC. 2017. 67–95.

11 Energizing the Future of Digital Technology in the Oil and Gas Industry

11.1 INDUSTRY TRENDS

It is emerging that the oil and gas industry has started moving toward a future of increasing digitization application. Companies investing heavily in developing these technologies will surely reap the benefits in years to come.

Digital platform will become the new business model and accelerate the reach of new markets. Artificial intelligence (AI) is already having a significant influence on current technologies. A cloud-based digital program that statistically predicts the future production of oil and gas is being adopted. Using its type curve studio, a user can forecast 500 oil well decline curves a minute after downloading from the database. Doing that by hand would take 2–3 minutes per oil well, or about 16–42 hours. In seismic shooting, Q-Marine survey geophysical data processing can generate a subsurface structural map in a matter of a few hours onboard a ship, which can better visualize the possible subsurface petroleum system. Cultivating a broader knowledge base gives engineers and geoscientists a wider perspective of a possible solution by advancing digital technology. Even an experienced and expert drilling engineer needs to keep track of digital technology. It is also essential for an oil field production engineer to better comprehend the exploration process and scenario to prepare a voucher for digitally selling petroleum products.

11.2 SCOPE OF EXPANSION OF DIGITAL TECHNOLOGY IN THE OIL AND GAS INDUSTRY

There is considerable scope for expanding digital technology in the oil and gas sector, starting from machine learning to robotics application. The oil and gas industry downturn, which commenced in 2014, has been a painful experience for many exploration and production companies. The first real signs of recovery came during 2016. During the last few years, slowly but surely, confidence, growth, and, perhaps most importantly, stability, etc., have all returned to the industry. Oil prices have stabilized and streamlined the industry. Companies are prepared for a "lower for the more extended" environment and continually innovate and develop technologies to increase efficiency. During the last few years, the development in digital technologies is beginning to change the industry over the next decade. As digital oilfields expand horizontally to encompass every aspect of operations and engineering, they will also expand vertically to touch

DOI: 10.1201/9781003307723-11

every functional discipline, accounting, finance, and executive management. Digital oilfields will become automated oilfields. Within that will be automated companies, with all information about the acquisition, development, production, and disposition of oil and gas assets managed in a centrally administered system with Business Process Management (BPM) processes, orchestrated workflows, and notifications. Changes to a fuel production plan in one support will be through the design of increasingly more sophisticated workflows that include economic analysis and roll-up to each revised portfolio optimization plan maintained by the finance department, with changes in expected Net Present Value being made immediately available for executive decision-makers.

11.3 GLOBAL GAS MARKETS

As India prepares itself to leap into the gas business, it would be worthwhile to draw inspiration from other countries that have maintained a large share of gas in their energy baskets. While markets like the United States and Europe boast vast domestic production and high consumption, gas trading hubs have been a catalyst in expanding the gas industry in these regions. Henry Hub of the United States and Europe's National Balancing Point and Title Transfer Facility had become benchmark hubs for the entire world. These markets have benefited from open access to infrastructure, system operator, unbundled marketing and transport functions, market-friendly transport access and tariff, and vital pipeline infrastructure. Most importantly, the Government will ensure favorable regulations, a clear example being unbundling of vertically integrated gas companies to make space for new participants.

11.3.1 Gas Markets in India's Context

Globally, the natural gas market has been gaining traction as a critical alternative and an ideal fuel to support the energy shift in favor of cleaner and greener energy resources. India, too, on its path to becoming a sustainable economy, also has set a vision to become a gas-based economy. The main aim is to increase the use of natural gas in the energy sector from the present value of 6%–15% by 2030.

India, too, has entered the group of such countries with its first delivery-based gas trading exchange. While it has just started its journey, it will take some time to realize the actual effect of such a trading platform, and the prospect is very high. Gas markets will play a pivotal role in achieving the national aspirations of increasing the contribution of gas. Such a type of platform will aid in expanding the gas-based industry by providing competitive and transparent pricing mechanisms, flexibility in procurement, and payment security. Ensuring price benchmarks will drive competition across the value chain and stimulate investments in exploration and production and downstream infrastructure. Moreover, market-driven and affordable pricing would boost industrial growth and economic competition.

11.3.2 Key Enablers to Evolve the Indian Gas Business

The Indian gas industry is experiencing exponential growth and is moving towards a market-based mechanism. It is essential to support its policy and regulatory framework to let the markets bloom. The Government's announcement of transmission tariffs and regulations by Petroleum and Natural Gas Regulatory Board (PNGRB) for operating gas exchanges will pave a growth path for the gas markets. This is in fundamental transformation; decoupling of marketing or selling of gas and its transportation is essential to help deter the formation of monopolies while ensuring greater competition, better utilization of pipeline infrastructure, and lower operating costs for end customers. Given that gas trade by nature offers the scope to flout agreed-upon contracts due to overdrawing or under-drawing, the launching of an overall controlling entity would bring more security and trustworthiness. An individualistic system operator would help ensure that the interests of all participants are protected.

Another critical aspect would be the more deregulation of pricing for domestically produced gas. This will allow more freedom to fix the price and market domestic gas and, in turn, boost domestic gas production, making it more viable for players to invest. This is the right time to integrate both domestic and Regasified Liquid Natural Gas (RLNG) markets since prices have converged for both. Moreover, inclusion under GST and an overarching regulatory framework will also play an essential role in boosting the overall gas markets.

11.3.3 Global Natural Gas Demand

Presently, global gas consumption is around 3,900 billion cubic meters (BCM). As per the BP energy outlook, the share of natural gas in the global energy basket will increase from 21% to 26%. In terms of volume, the demand is likely to increase from 3,900 to 5,200 BCM. In the business-as-usual (BAU) scenario, gas demand increases throughout the next 30 years, increasing by one-third to around 5,200 BCM by 2050. This growing demand in gas consumption is relatively widespread, with solid increases across developing Asia, Africa, and the Middle East. The outlook for gas is assisted by broad-based demand and the increasing availability of global gas supplies. The global market varies significantly across the scenarios. The significant combination of low emissions and high flexibility positions natural gas as a crucial energy source to meet the global energy demand while establishing a solid transition path to a net-zero energy system:

- Supporting a shift away from conventional coal in fast-growing, developing economies like India where renewable energy and other non-fossil fuels may not grow sufficiently quickly to increase and replace coal faster. The total proven gas reserves are 198.8 trillion cubic meters (TCM). Based on the present gas production to replacement ratio, gas production can last up to 49.8 BCM years to be precise, based on the existing reserves.

11.4 THE WAY FORWARD IN THE GAS BUSINESS

India currently has a total oil and gas pipeline of 36,300 km and an extensive gas pipeline infrastructure of around 17,700 km and plans to double the total pipeline capacity in 2–3 years. Coupled with necessary policy and regulatory modification, India can indeed accelerate its pace toward building a market-based gas market economy.

As also suggested by International Energy Agency, a vibrant, dynamic, and well-functioning domestic gas market would be a strong pillar for building a gas economy in India. A well-functioning market requires enabling the regulatory framework to reduce entry/exit barriers to transactions, achieving a balance between safeguarding the interests of best customers pricing, delivery expectations, and seller's returns expectations.

Bringing in measures to build vibrant and competitive gas markets is necessary. A dynamic market will, in turn, ensure the exponential growth of the industry by fueling consumer demand and proliferation of supply. These endeavors will implement a pipeline, Supervisory Control and Data Acquisition (SCADA) application distribution, monitoring, and control of gas flow networks.

11.5 ENERGY PERSPECTIVE

Energy assurance in totality needs many solutions between the present availability of fossil fuel and contribution from non-conventional energy sources like solar power and wind energy contribution in present-day energy prospective in India. The gift of indigenous oil is 18% of consumption, whereas non-conventional energy with an accelerated impletion effort can only reach 10% of the required energy basket. The gas market is also moving faster concerning availability by implementing gas-related work like piped natural gas (PNG), compressed natural gas (CNG), and city gas distribution (CGD) networks after commissioning of new LNG plants. The International Energy Agency (IEA) performs as a policy adviser to its member states and significant emerging economies such as a few developing countries like Brazil, China, India, Indonesia, and South Africa to assist energy security and advance the clean energy transformation worldwide. The mandate for IEA has broadened to focus on providing data, analysis, policy recommendations, and solutions to help countries secure affordable and sustainable energy. In particular, it has concentrated on supporting global efforts to accelerate the clean energy transition and mitigate climate change.

For the quick innovation of new technology both indigenously and imported source, there needs to be a balance policy between the corporate sector and government participation, e.g., solar energy and wind energy are available on a 24×7 basis, which include the need of storage technology for about 12 hours. To reduce the dependency on diesel and petrol in the auto industry, electric vehicle (EV) is a new technology that is on the horizon. The essential component of ending market value (EMV) is the battery and establishment of suitable charging infrastructures, which many consider being developed indigenously. The cost of lithium batteries is incrementally reducing in the western world; India may consider this part of innovation for private participation. The Government may encourage battery development indirectly through IITs and other R&D institutes, and NITI Aayog may monitor its progress and coordinate the projects being developed by institutes.

11.6 INITIATIVES IN THE EXPLORATION AND PRODUCTION SECTOR BY THE GOVERNMENT OF INDIA

Hydrocarbon industry can look forward to further liberalized and forward-looking policies with the Minister for Petroleum & Natural Gas, Government of India. The

Government took initiatives to usher in policies like Hydrocarbon Exploration and Licensing Policy (HELP), Open Acreage Licensing Policy (OALP), previously held nine NELP Rounds, Small Discovered Fields and Exploration & Exploitation of Unconventional Hydrocarbons under Existing Production Sharing Contracts, Coal Bed Methane Contracts, etc., which were rolled out meticulously. More such aggressive policies will further show positive results in increased oil and gas production since the upstream sector's introduction of HELP and OLAP regimes. The Government of India has initiated a policy framework for incentivizing Enhanced Oil Recovery/Improved Oil Recovery. To further incentivize the production from domestic gas resources, marketing and pricing freedom has been introduced to those new gas discoveries whose Field Development Plans are yet to be approved. Fiscal incentive is also provided on additional gas production from domestic fields over and above-average production.

In India, with shallow water contributing to 60% of the country's oil and gas, deepwater development is also going on for quite a few years, spearheaded by PSU & Private operators on the East Coast of India. In 2021, despite a few initial setbacks, both ONGC and RIL are producing a significant amount of gas from deepwater blocks. India has a deepwater potential of over 10 billion tons of prognosticated reserves of oil and gas equivalent in its 1.35 million km^2 deepwater acreages. deepwater acreages account for 39% of oil and gas acreages. India has 26 sedimentary basins measuring a total of 3.14 million km^2. West Coast, East Coast, and Andaman Deep Offshore are the major deepwater basins with colossal potential yet significant developments. Shallow water development and production is mainly prevalent in the Western Offshore of India. Given this growth and developmental backdrop and the vast shallow and deepwater potential, India offers a huge opportunity to do business with. The Government of India has been quite encouraging and has rolled out many industry-friendly measures to fast track the efforts through recent reforms like OALP & differential scanning fluorimeter (DSF) rounds. Since 2016 DSF rounds I & II and OALP rounds, I, II, III & IV, a total of 35 blocks have been awarded to different operators.

Deepwater development faces many challenges such as higher degree of expenditure, huge risk, and technology intervention that need special attention, and customization of tools, technologies, and approaches specific to fields is also essential.

11.6.1 Future Outlook of India's E&P Industry

It is known that India imports 84% of its crude oil need. Government is keen on reducing import dependency. Therefore, India's domestic oil production will play a significant role in meeting the oil demand. India has an estimated sedimentary basin area of 3.36 million km^2 and consists of 26 sedimentary basins covering 1.63 million km^2. The site is on-land, shallow offshore up to 400 m isobaths with an area of 0.41 million km^2. Deepwater beyond 400 m isobaths has a sedimentary area of 1.32 million km^2. According to the recent Directorate General of Hydrocarbons (DGH) estimates, 26 basin have in-place hydrocarbon resources of 41.872 billion tons of oil-equivalent, of which 29.796 billion tons are undiscovered hydrocarbons.

11.7 TECHNOLOGICAL WAY FORWARD

Some of the upcoming technological initiatives for both E&P business will be deliberated in the following subsections.

11.7.1 Real-Time Virtual Flow Metering (RTVFM)

RTVFM is new technology for the measurement of flow in oil, gas and condensate field. Reliable estimation of production flow rates for wells is of paramount importance for 3D model history matching, well-test interpretation, back allocation, real-time monitoring, and reservoir management. Virtual metering technology can evaluate production rates well and is based on real-time online data and analytical models.

RTVFM technology had been developed by integrating a software platform and mathematical model. The algorithms simultaneously solve dynamic pressure and temperature gradients and the bottom hole choke equation to find optimal solution rates that match physical readings. It had an option that if one monitor fails, other takes over. This technological lift, i.e., a virtual metering workflow network in the real-time oil and gas production calculation, implements the digital oil field as shown in Figure 11.1.

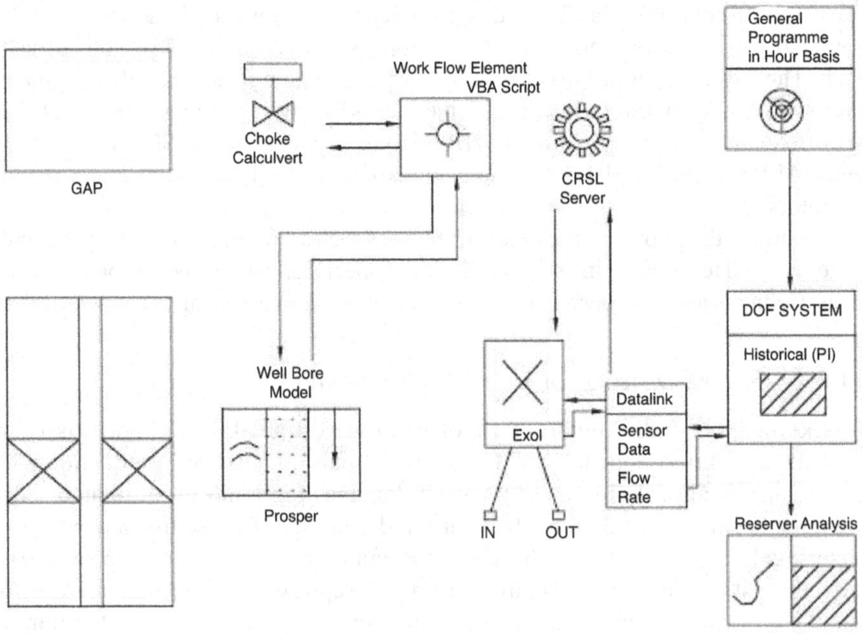

FIGURE 11.1 RTVFM flow diagram.

11.7.2 New-Generation Production Logging

The first-generation production logging tool (PLT) provided a single, discrete measurement for each sensor along the tool assembly's length, resulting in long-length tool assemblies and measurements taken at different points along the flow path. This approach had many drawbacks: long tool strings, point sensors that only provided a measurement at a single point in the cross-section of flow, and measurements not acquired simultaneously at each depth of logging. Second-generation PLT represented an improvement by arranging the sensors in an array, enabling multiple measurements to be made at a single depth. However, the tool strings were still long, and not all were arranged optimally to capture data in the flow path. Third-generation PLT is one-tenth the length of the first-generation and about one-third of the second-generation tool in size. The digitalization process allows for direct measurement of flow conditions and quick processing and interpretation of results. The multiphase flow and deviated wells, co-locating sensors in a spatial geometry, provide the optimal information to create an entirely accurate flow behavior in the downhole flow.

Third-generation PLT is a miniature and digitized version of traditional equipment in terms of efficiency and capability. The tool has, within a 3 ft. length, up to 24 sensors that collect multiple measurements of fluid properties and fluid movements in the wellbore, i.e., oil, gas, and water hold-up and bubble count, liquid conductivity, phase velocities, rotation, pressure, temperature, rates, and depth correlation plus power and communication. The fluid characteristics are locally screened by an array of 8–16 tube-shaped probe sensors that are interchangeable, depending on target depth.

11.7.3 High-Fidelity Seismic Imaging for Reliable Mapping of Complex Reservoirs

The Q-marine point-receiver marine seismic system was launched in 2000 and introduced several new concepts to the towed-streamer market. The method set the benchmark for other acquisition technologies and continues to perform at today's highest levels. Based on an integrated system consisting of the recording (TriAcq), positioning (TriNav), and source (Tricor) subsystems, it features a unique solid gel-filled, zero-spill, point-receiver streamer. Unlike other systems, all in-sea components such as acoustic positioning and streamer control devices (Q-Fins) are integral to the streamer, which reduces noise and drag. Since its introduction, the system has successfully delivered across various basins and environments worldwide. At all stages across the E&P cycle, the Q-marine system achieves value through high resolution and repeatability within the required time frames.

11.7.3.1 Point-Receiver Measurements

A significant source of noise that affects marine seismic data quality and repeatability is related to the movement of the streamer through the water. This is accentuated by adverse weather and sea conditions. The Q-marine system uses single-sensor recording to characterize noise with sufficient spatial fidelity, eliminating it using targeted digital filtering techniques. This does not influence the signal bandwidth or fidelity, unlike conventional noise suppression.

11.7.3.2 Steerable Streamers

The Q-marine system pioneered steerable streamers by introducing the Q-Fin marine seismic streamer steering system. The Q-Fin system's remote-controlled wings reduce noise and enable precise depth control and horizontal steering. Horizontal streamer steering provides feather correction, safe streamer separation control, fan-mode shooting, active steering for optimal coverage, and 4D repeatability.

11.7.3.3 Calibrated Sources and Positioning

The signature of any seismic source varies from shot to shot for many reasons, including source dropouts and variations in array geometry. This variation reduces the accuracy and repeatability of seismic data. The Q-marine system has an advanced digital source controller that enables calibrated marine source—a fully calibrated source signature derived for every shot utilizing the notional source method.

Q-marine is a 3D-seismic data acquisition system for visualization, processing, and interpreting subsurface geological structures for modern-day seismic shooting.

11.7.4 3D Virtual Modeling and Drone Technology

Drone technology and geological strata outcrop modeling have already changed the way future fields are party-wise conducted and provided an invaluable new learning resource for geoscientists. Other emerging digital technologies that have and will continue to affect the oil and gas industry operators are connected with 3D virtual modeling and drone technology.

In the case of implementation of drone technology, this could have considerable benefits in terms of monitoring and inspection of inaccessible oil and gas surface facilities. With drones, visual inspections of such onshore facilities can be carried out remotely, with high-quality images fed back in real time to drone operators. Multiple number of facilities can be inspected within a few hours rather than potentially delay inspections being carried out in person. This type of drone inspection can then recognize actual issues and problems and direct the operational staff to deal with them immediately, improving efficiency by focusing on those facilities that require immediate attention. Another example is the inspection/maintenance of gas flare platforms offshore. Difficult to reach and complex, this type of physical inspection can require oil and gas production shut down before it can take place, but with drone technology, this inspection can be conducted at any time, decreasing expenditures and increasing safety of the installation.

In another application, drones can capture incredible images of rock outcrops for geological survey. When combined with other technologies such as airborne survey and laser scanning (light detection and ranging, LiDAR) data, it can be geo-referenced and processed to generate virtual outcrop models that benefit classroom learning. In addition, if these data can be obtained from the outcrop via physical inspection, these virtual models can be exemplarily tuned with the actual, high-resolution 3D digital model (land Q-Marine survey) of the outcrop, which can be visualized and interrogated by the user for mapping. Some possibilities of data and visualization from such models include extracting 3D lines, which can demonstrate sloping deposits, levees channel bodies and sedimentary structure, fracture and fault geometry, unconformities, and fancy boundaries. Additional data like porosity, permeability, and net to gross can be combined with information on the thickness and width of sedimentary bed features to arrive at a more powerful, interactive learning resource that can supplement more traditional classroom and field-based training season.

11.7.5 Machine Learning Based on the Early-Warning System

Major oil fields in India have entered a high degree of water-cut production stage after decades of water flooding; hence, stabilizing oil production will be very arduous. After repeated stimulation treatments throughout the field's history, abnormal decline rates—i.e., exceeding 5%—occurred more frequently. Production declines will be dramatically high.

An effective early-warning system (EWS) was incorporated, releasing a production alarm to enable engineers to take preventive measures. Factors that can affect the unusual decline were selected. The index sets of product composition and injected and produced water obtained from practical statistics were considered the leading assessment indicators. Grey relational analysis assessed the different indicators' importance and eliminated redundant parameters.

Machine learning was assumed to build the EWS. Using the degree of deviation from ordinary as the input data for the prediction model provided soaring accuracy. However, the basic machine-learning model consists of many input parameters that cannot be obtained easily. The number of input parameters was optimized based on the accuracy variation under different input parameter numbers. This will improve the forecast accuracy; artificial samples were added to the operator training process.

The prediction accuracy (some say) of the final optimization model can reach 92%. This result reveals the possibility of an anomalous decline in different reservoirs, which can guide oilfield production strategy effectively. The EWS is verified by an oilfield production expert.

Neural networks use the present data to determine the inferred model between input and output. Classifying and prediction are the most common exercise. Neural networks are utilized in the industry to resolve categorization and regression problems.

The back-propagation (BP) neural network is a multilevel feed-forward network that corrects the link weight of each level from back to front according to the contrast between the actual output and the likely output. The primary model consists of one input layer, a second output layer, and several hidden layers. This will also depict the nonlinear relationship between input variables and output variables. More training programs are needed for obtaining multiple samples from each layer's connection weight and other information as knowledge storage to predict new pieces. The concept of applying BP neural-network technology to the all-inclusive evaluation of specific problems is to take the assessment index system as the input vector and the quantity value representing the corresponding comprehensive evaluation as the output vector. First, the network is trained using samples that have succeeded through a traditional thorough assessment. After training and learning, the weight of each indicator can be indicated rightly, and the trained neural network can be used as a successful tool for total valuation. For the three-step EWS that will be discussed, the warning alert is divided into three levels based on a decline-rate variation. A decline rate of less than 3% is considered a normal condition. A decrease rate between 3% and 5% signals a medium-warning alert. A rate of decrease of more than 5% issues a high-warning watch.

11.7.6 Establishment of an Early-Warning Indicator System

Grey relational analysis was utilized to identify the correlation between the decline rate and the input parameter. Compared with conventional multifactor correlation and regression analysis methods, the grey correlational study requires fewer data and calculations, making it widely used. Figure out the steps in the analysis process and include various measures. Tabulate the correlation degree and rank among the oil production decline rate and various influencing factors.

11.7.7 New Oil and Gas Well Hydro-Fracturing Technology

New hydrofracturing techniques are developed with the discovery of shale gas exploitation process. These fractures are a very complex process, with dozens of parallel fractures in a thin thickness growing simultaneously during the execution of the job in the tight reservoir. The concept of fracture initiation had also changed with the development of new fracture diagnostic tools such as tilt meter fracture mapping and micro-seismic mapping, which can provide better fracture dimension and feed fracture propagation models. More extensive fractures need more lateral length to reach more area for better productivity. The number of proppants and fracturing fluids will increase to the manifold, i.e., the requirement of average proppants mass volume had risen to 3 times during the last couple of years, the condition of intermediate fracturing fluids had increased to 2.5 times during previous 3 years, average stage count had also increased to 3.5 times, and also average pumping rates had risen to 2.5 times during the last 3 years. Service companies have become more skilled, making redundant equipment and spare available at the site.

11.7.8 Reservoir Simulation Modeling

Earlier reservoir visualizations are more related to conventional oil and gas reservoirs. These reservoirs have a higher gross thickness, porosity, permeability, and high organic carbon content. In the petroleum system, source rocks are more mature and seem to be more driving force for a fluid movement from dry gas in very mature layers, progressing through gas condensate and volatile oil to black oil of less than 100 gas-oil ratio (GOR) (M3/M3). The basic assumption of reservoir modeling used to consider the history of the geographically close wells, i.e., the forecasts for new zones, involved numerical simulation of regionally distributed properties like reservoir fluid properties. The equation described fluids' pressure, volume, and temperature in the known zones used for exploitation. In the withdrawal of oil, pressure and compressibility are closely related. In the recovery process in a given reservoir, the pressure drop is directly proportional to compressibility. Compressibility is used as a significant parameter to qualify the model. These are used for three types of oil reservoirs for modeling: low-GOR black oil, high-GOR black oil, and volatile oil. All these fluids, volume corrections can be done by the Peng–Robinson method, but this method had a limitation for representing oil-volume changes with pressure. A difference as low as 5% relative volume can generate 30% difference incompressibility. So compositional model simulation is the best approach for volatile oils where mass transfer between the oil and gas is insignificant. When production forecast from simulation model shows difference, a 5% difference in final recovery results up to 30%.

11.7.9 Transformation of Offshore Analog Fields to the Digital Domain by AI

Monitoring daily oil and gas production data like flow, pressure, and temperature of each well is necessary, but this needs further implication of expenditure. It is a luxury to have a multiphase meter for individual well during well testing. Instead, most fields need to share metering facilities using a test manifold. Wells are then tested with a specific frequency with the help of a well-test separator, and the daily rate is estimated based on these test results. Generally, well production between tests is assumed to be constant until a new test is conducted. This assumption can be improved by using a calculated value from a virtual flow meter (VFM). The input parameters are collected from the wellhead pressure gauge. Wellhead pressure is measured from analog gauges; with the help of a pressure transmitter (I to P convertor), the data can be transferred through a SCADA network, which is read in the central control room in digital form, and this way human errors can be eliminated. An intelligent well system is provided with bottom pressure transmitter, flow transmitter, temperature transmitter, etc., which are linked up to surface SCDA system, and well test manifold may be equipped with a multiphase meter or VFM.

11.8 DIGITAL APPLICATION IN THE OIL REFINERY SYSTEM

The oil refinery process system had incorporated different types of software system for day-to-day operation and planning purposes. In India, the 30%–32% refinery adopted additional software system like Integrated Manufacturing Operations Management for better management of refinery management. Similarly, refinery plant operations use software for tasks such as pre-heater train, heat exchanger fouling prediction, product blending performance analytics, and predictive asset maintenance systems.Other digital service implementations are Digital Unified CRM (Retail) as-a-Service Dynamic Pricing, Integrated Project Controls & Analytics Connected to Men, Machines, Materials Drone-based surveys, AR/VR-driven Working Methods Capital Projects, Suite INFRASTRUCTURE Supply Chain Control Tower Commercial Optimization Fleet, and lastly Tracking & Optimization Terminal Automation.

11.9 MIDSTREAM MANAGEMENT

In the retail industry, digital solutions allow oil companies to enhance their operational efficiencies, exceed customer expectations, and gain a larger market share. For instance, a government-operated oil and gas company uses the Fuel Retail Operations Platform and the Fuel Retail Insight Solution to gain a competitive edge by accessing actionable, near real-time micro-market field-level insights into its operations and overall retail business.

11.10 PIPELINE, TRADING, AND PROCUREMENT

There is a new dimension in digital application in the Midstream Industry. When laying new cross-country pipelines, which has become more of a country base issue, due in no small amount to fears over leaks, a new fiber optic-based pipeline detection, monitoring technology had arrived at an appropriate time in pipeline sectors. Hifi Engineering Inc. has developed a high-fidelity dynamic sensing system furnished with a fully distributed monitoring system using a specialized fiber optic cable designed to detect high-fidelity acoustics signals, temperature, and vibration/stain. In addition to supplying high-confidence leak detection across every centimeter of the pipeline networks in real time, preventative conditions can also be detected, such as third-party interference, machinery operating too close to the channel, thermal anomalies, or geotechnical events such as earthquakes and landslides being monitored regularly.

HIFI recently endorsed the ability to retrofit the technology to existing pipelines with a pilot. TransCanada Corporation had introduced a section of the Keystone pipeline into Houston, USA. The companies cooperate along with Alberta Innovates and the Alberta Small Business Innovation and Research Initiative program to develop a real-time verification and diagnostic system to ensure the ongoing operation of the technology remained finely tuned and effectively operated. The company's exceedingly sensitive fit-for-purpose fiber optic cables generate terabytes (1 trillion bytes) of data every day—so keen this can discriminate between a full and an empty dump truck driving on the road above a buried pipeline or an earthquake originating thousands of kilometers away. From this richness of data, the real challenge is figuring out a concerning event and routine happenings. False positives are the scourge of most monitoring systems.

Intelligent signal processing algorithms must transform the fiber optic sensing data into actionable information. A Hifi–University of California recent partnership is mainly focused on harnessing the power of AI and deep learning technology to furnish the highest level of safety and security for pipeline operations.

Life has had a strong relationship with the University's Schulich School of Engineering over the past number of years, including collaboration on developing the company's cutting edge leak detection algorithms, added by Mr. Ehsan Jalilian, Hifi vice-president, R&D, himself a University of California graduate. Directed the oil and gas industry's urgent need for enhanced pipeline safety, he said this industrial–academic partnership has potential national and global significance.

"All this new cutting edge research that are doing with the University of California has been, an integral component of our success," said Mr. Jalilian. The company did hire a few super-talented team of corporate data scientists and signal processing engineers from the university who had to work full time with them. Out of a dozen engineers, probably nine or ten are University of California graduates on the systems team.

The new types of pipeline detection software are like IMM-2.0 PIDS (Pipeline Intrusion Detection System), Pipeline LDS (Leak Detection System), Corrosion Management of Intelligent Pipeline, and other various types of commercial analytical software such as Spend Analytics, Payment Terms Analytics, Cost Modeling Inventory Optimization, and Market Intelligence of Bid Analytics. The above software is available in the market, which will be helpful for better digitization. India had also successfully embraced digital technology for integrated pipeline management.

11.10.1 Fueling Efficiency in Upstream

The innovation in the digital oil and gas processing industry, like lubes, needs to address the growing demand and improve the costs while future-proofing their business in a competitive market. This will get from lowering raw materials, which count for more than two-thirds of the total production cost. High-end analytics models may help achieve this by using base oil and additives, lowering tolerance limits in manufacturing blending plants, and increasing the bottom-line value. Also, access to real-time data may help companies dramatically reduce high inventory and logistics costs.

11.10.2 Future HR Digital Consumer

HR digital solutions such as providing predictions of certain events, video analytics, digging into data with drill-downs, and integration tools now provide companies clearer visibility into projects and help detect problems early on and take quick action. The benefits may be a 10%–15% improvement in project schedules, a 5%–10% increase in asset productivity, and more than a 5% reduction in cost. Cutting costs and improving recovery rates are vital priorities for the upstream sector. Sometimes, a lack of coordination between multiple contractors in managing capital projects leads to significant delays. Digital solutions offer some profound benefits. Modern solutions such as Digital Asset Maintenance and Production Surveillance are helping to increase the efficiency further in oil and gas E&P companies and minimize downtime through predictive maintenance. Examples of human resource management software include Hiring & Retention Analytics, Performance & Talent management, and Employee Engagement Learning & Qualification Analytics.

11.11 ELECTRIC VEHICLES

The threat of perception from EVs also looms large as the Indian Government resolves to promote such vehicles and starts working to have better technology for manufacturing low-cost lithium batteries. The winners will reinvent themselves by adopting new and agile business models encouraged by digital technology. Those who utilize the power of technology to differentiate themselves and find new ways to innovate and get bigger will gain an important initiative.

Many technologies are available in conventional vehicles; plug-in electric cars (also known as electric cars or EVs) have different capabilities to accommodate additional drivers' needs. A significant feature of EVs is that drivers can plug them in to charge from an off-board electric power source. This differentiates them from hybrid electric vehicles (HEVs), which supplement an internal combustion engine with battery power but cannot be plugged in.

There are many basic types of EV: All-electric vehicles (AEVs) and plug-in hybrid electric vehicles (PHEVs). AEVs include battery electric vehicles (BEVs) and fuel cell EVs. In addition to charging from the electrical grid, both types are partially charged by regenerative braking, which generates electricity from some of the energy usually lost when braking. Which type of vehicle will fit your lifestyle depends on your needs and driving habits. Find out which BEVs and PHEVs are available to suit your needs.

Electric cars can use AC or DC motors: If it is a DC motor, it may run on anything from 96 to 192 V. If it is an AC motor, it probably is a three-phase AC motor running at 240 V AC with a 300 V battery pack. Most PHEVs and AEVs use lithium-ion batteries like these. Energy storage systems and standard batteries are essential for HEVs, PHEVs, and AEVs.

There are three main types of EVs, classed by the degree that electricity is used as their energy source:

- HEVs, which are powered by petrol and electricity
- PHEVs
- BEVs

AEVs run only on electricity. Most have 80–100 miles of all-electric ranges, while a few luxury models range up to 250 miles. When the battery is depleted, it can take from 30 minutes (with fast charging) up to nearly a full day (with Level 1 charging) to recharge it, depending on the type of charger and battery.

If this range is not sufficient, a PHEV may be a better choice. PHEVs run on electricity for shorter spans (6–40 miles) and then switch to an internal combustion engine running on gasoline when the battery is depleted. The flexibility of PHEVs allows drivers to use electricity as often as possible while also fueling up with gasoline if needed. Powering the vehicle with electricity from the grid reduces fuel costs, cuts petroleum consumption, and reduces tailpipe emissions compared with conventional cars. When driving distances are longer than the all-electric range, PHEVs act like HEVs, consuming less fuel and producing fewer emissions than similar conventional vehicles.

The basic main components of EV are as follows:

- Traction Battery Pack (A): The function of the battery in an electric car is as an electrical energy storage system in the form of direct-current electricity (DC).
- Power INVERTER (B).
- Controller (C).
- Electric Traction Motor (D).

BEV, pure EV, only-electric vehicle, or AEV is a type of EV that exclusively uses chemical energy stored in rechargeable battery packs, with no secondary source of propulsion (e.g., hydrogen fuel cell and internal combustion engine). A battery-operated EV is explained in Figure 11.2.

Battery electric vehicles, or BEVs, use electricity stored in a battery pack to power an electric motor and turn the wheels. When depleted, the batteries are recharged using grid electricity, either from a wall socket or a dedicated charging unit. The amount of pollution produced depends on how the electricity is made.

FIGURE 11.2 Electric vehicle 3D diagram.

It turns out most electric car drivers don't even bother to plug in every night or necessarily to fully charge. People have regular driving habits, and if that means 40 or 50 miles a day, a couple of plug-ins a week is fine. Self-driving cars are ready to revolutionize the transportation industry. At the heart of autonomous technology is LiDAR, a vehicle vision system that measures distance by illuminating a target using light and lasers as its primary sensor; automotive LiDAR sensors allow driverless cars to see. Some researchers claim that, without LiDAR's capabilities, self-driving vehicles could not progress much further than what the Society of Automotive Engineers terms autonomy level 3 abilities. Numerous manufacturers have integrated or invested in LiDAR. This eBook, *LiDAR – Future Technology for Autonomous Vehicles*, provides an in-depth description of LiDAR technology and its applications. It also lists the leading LiDAR vendors worldwide and looks at what vehicle manufacturers have invested in or integrated LiDAR. Furthermore, this eBook offers insights into the automotive LiDAR industry's direction. Apart from going through the basics of LiDAR, the eBook *LiDAR – Future Technology for Autonomous Vehicles* provides a better understanding than other technologies:

- LiDAR will be cheaper—but it's still too expensive for production vehicles.
- Top LiDAR companies in the automotive industry.
- Snapshot: Japanese LiDAR supplier Pioneer.

11.12 DIGITAL POWER-DRIVEN CONSUMER-CENTRIC MARKETING

Advanced analytics modeling serves as a leading global lubricants player to improve visibility into its B2C supply chain and reduce distributors' much-needed working capital requirement in India. With insights into the market potential of each of the 400 districts in India, the oil marketing company had planned its sales and marketing activities superior. The company is also increasing its sales force effectiveness with AI solutions such as News Page. In effect, it transitions to a new business model—of selling services and lubricants. Digital tools have also helped optimize the warehouse network, reducing 8%–10% logistics costs.

11.13 DOWNSTREAM APPLICATIONS

AI-related technologies in oil and gas do not stop with exploration and production. Many operators in the petrochemical and refining sector rely on predictive analytics and model predictive control to continuously improve the overall performance of their facilities and more effectively maintain the condition of the equipment.

Many companies are even applying it to streamline oil and gas transportation, refining, and distribution. These firms utilize advanced algorithms to analyze data such as economic conditions and weather patterns to forecast consumers' demand, which allows for better allocation of resources and optimal pricing.

EVs sales, excluding e-rickshaws, grew by 20% in India in 2019–2020, as per report by industry body Society of Manufacturers of Electric Vehicles (SMEV). In 2019–2020, a total of 1.56 lakh EVs were sold in the country against 1.3 lakh units in the previous fiscal year, SMEV said.

Globally, the EV market is growing at a fast pace. The Indian Government's vision is to convert 58% of all public transport and new cars to EVs by 2030. Global energy report by BNEF forecasted that EVs will reach 10% of global passenger vehicle sales in 2025, with that number rising to 28% in 2030 and 58% in 2040.

According to the study, EVs currently occupy 3% of the global car sales market. The Government will need to move beyond availing subsidies to automakers and consumers procuring EVs and focus on building an authorizing ecosystem. Digital technologies are already serving to create new business models for electric cars, for example, battery swapping, imposing infrastructure management, and fleet management models. Digitalization shows immense perspective to empower oil and gas companies to pursue new transformation and growth at speed and scale. But, to truly seize the chances in digital offers, companies need to take a broader approach. Establishing up a digital Center of Excellence to drive the highest worth from related resourcefulness must be part of their digital transformation journey.

11.14 FINAL LINK IN THE DIGITAL CHAIN

Recently, the Calgary-based company introduced the final link in the chain for digital end-to-end billings management between oil and gas companies and service vendors of the oil and gas business. That link is a new cloud-hosted vendor invoice portal, which ties directly into invoicing and coding appreciation and JV accounting and financial reporting to provide a seamless invoice-to-pay workflow solution. By the end of 2018, ten producers had implemented it into their payables workflows, and more vendors came onboard to submit their invoices in the coming years.

Previously, producers and vendors grappled to connect their data through either ineffective, manual, paper-based billing processes or cobbled-together, multi-source solutions that failed to smooth the workflow for either of them. Internal and external barriers hamper workflow, resulting in incomplete or incorrect data transferring from vendors to accounts payable groups. Invoices received through various channels and templates required ample time to consolidate, leading to errors, coding discrepancies, and invoices being lost entirely.

Mr. Pandell's analysis of 50 Alberta-based service companies using a paper invoicing process showed an average invoice-to-pay cycle of greater than 60 days.

The preferable approach for managing a large number of users is when there are over 400 users. Mr. Pandell was aware of where to find the solution. The accounting software collects quarterly feedback during the company's Client Advisory Board sessions.

This section identified two digital workflow opportunities that could lower industry challenges and enhance the invoice-to-pay process—one between the producers' accounts payable and financial accounting groups and the second between their payable groups and service providers. The first was solved through a cloud-based account-payable structure that automates invoices through a configurable approval process, and the second one was solved with a web portal that expands AP and JV's financial validation tools, audit controls, and digital workflow to the supply chain in real time.

11.15 VALUE CONSOLIDATION

Consolidate digital-specific capabilities to leverage the right talent and resources across the organization as and when required. Experts recommend that oil and gas companies in India act on these five imperatives to achieve digital's total value: few innovations versus several pilots. Instead of undertaking several digital advantages, pick a handful across all aspects of the business. Technology will keep swapping, but companies need to place their bets.

11.15.1 Rethink, Redefine, and Redesign Process

When accounting digital technologies for any stage—exploration, development, production, midstream, refining, or marketing—be clear about the intended future state, constraints, and occasions.

11.15.2 Make Change Stick

Couple of digital companies put money with investments in people and culture. Invest in tools and training to develop an environment of participation and experimentation.

Analytics help to understand how causal relationships influence decision-making and organizational agility. Creation of small teams can work at speed to drive digital mode in the company. The modern trend is economic liberation. Now, it is cryptocurrency posing a threat to financial stability. The banking sector will face a problem because cryptocurrency becomes an alternate bank deposit and banks lose the deposit; their ability to create credit gets constrained. This has a grave implication, both for the power of monetary policy to influence interest rates and for economic growth in a bank-driven economy like India already struggling with low credit offtake. That's not all. A migration of mining activity to growing market economics can have severe implications for capital flows and energy consumption. Increased demand for crypto assets could facilitate capital outflow that affects the foreign exchange market. Crypto exchanges may play the crucial role of facilitating local currency conversion to crypto assets and vice versa.

11.15.3 Rethink, Redefine, and Redesign

When considering digital technologies for any stage—exploration, development, production, midstream, refining, or marketing—be clear about the desired future state, constraints, and opportunities.

11.16 PROGRAMMING LANGUAGES

Programming languages played a major role in the evolution of AI since the late 1950s, and several teams carried out important research projects in AI, e.g., automatic demonstration programs and game programs (Chess, Ladies). During these periods, researchers found that one of the special requirements for AI is the ability to easily manipulate symbols and lists of symbols rather than processing numbers or strings of characters. Since the languages of the time did not offer such facilities, a researcher from MIT, Mr. John McCarthy, developed, during 1956–1958, the definition of an ad-hoc language for logic programming, called LISP (List Processing language). Since then, several hundred derivative languages, so-called Lisp dialects, have emerged (Scheme, Common Lisp, Clojure). Indeed, writing a LISP interpreter is not a hard task for a LISP programmer (it involves only a few thousand instructions) compared to the development of a compiler for a classical language (which requires several tens of thousands of instructions). Because of its expressiveness and flexibility, LISP was very successful in the AI community until the 1990s.

Another important event at the beginning of AI was the creation of a language with the main purpose of expressing logic rules and axioms. Around 1972, a new language was created by Mr. Alain Colmerauer and Mr. Philippe Roussel named programming language where the expected logical rules of a solution can be defined and the compiler automatically transforms it into a sequence of instructions.

Prolog is used in AI and in natural language processing. Its rules of syntax and its semantics are simple and considered accessible to non-programmers. One of the objectives was to provide a tool for linguistics that was compatible with computer science. In the 1990s, the machine languages with C/C ++ and Fortran gained popularity and eclipsed the use of LISP and Prolog. Greater emphasis was placed on creating functions and libraries for scientific computation on these platforms, and they were used for intensive data analysis tasks or AI with early robots. In the middle of the 1990s, the company Sun Microsystems started a project to create a language that solved security flaws, distributed programming, and multi-threading of C++. In addition, they wanted a platform that could be ported to any type of device or platform. In 1995, they presented Java, which took the concept of object orientation much further than C++. Equally, one of the most important additions to Java was the Java Virtual Machine (JVM) which enabled the capability of running the same code in any device regardless of their internal technology and without the need of pre-compiling for every platform. This added new advantages to the field of AI that were introduced in devices such as cloud servers and embedded computers. Another important feature of Java was that it also offered one of the first frameworks, with specific tools for the internet, bringing the possibility of running applications in the form of Java applets and Java scripts (i.e., self-executing programs) without the need of installation. This had an enormous impact in the field of AI and set the foundation in the fields of web 2.0/3.0 and the internet of things.

However, the development of AI using purely procedural languages was costly, time-consuming, and error prone. Consequently, this turned the attention into other multi-paradigm languages that could combine features from functional and procedural object-oriented languages of various software providers, and their Logos are listed together for easy handling in list of programming languages provided in Table 11.1.

TABLE 11.1
Programming languages with Logo

Logo	Language	Date	Type	Influenced By	AI Resources
	Lisp	1958	Multi-paradigm (functional, procedural)	IPL	• Homo-iconic easy to deal with large amount of data • Good mathematical alignment • Lots of resources for symbolic AI (Eurisko or CYC)
	C++	1983	Procedural	C, Ag Cl 680	• Fast execution times • Some compatible libraries for AI such as Alchemy for Markov logic and Mlpack for general Machine Learning (ML)
	C#	2000	Multi-paradigm (functional, procedural)	C++, Java, Haskell	• Easy prototyping and well-elaborated environment • One of the most used languages for AI gaming and provides good compatibility with popular game engines such as Unity
	Clojure	2007	Functional	LISP, Erlang, Prolog	• Easy design and cloud infrastructure that works on top of the JVM • Rapid interactive development and libraries for development of behavior trees (alter-ego)
	Erlang	1986	Functional, concurrent	LISP, Prolog	• Good framework to deal with concurrency and elastic cloud (scalability) • Libraries for logic programming such as Erlog
	Go	2009	Procedural, concurrent	Algo, CSP Python	• Easy concurrency and asynchronous patterns with a decent runtime • A few libraries for machine learning such as Golearn
	Haskell	1990	Functional	Lisp	• Easy parallelization and possibility of handling infinite computations • A few utilities to implement neutral networks (LambdaNet) and general ML (HLearn)
	Java	1995	Procedural, concurrent	C++ Ada 83	• Virtual machine provides efficient maintainability, portability, and transparency • A myriad for libraries and tools for AI such as Tweety and ML (WekaDeeplearning4j, Mallet, etc.)
	Julia	2012	Multi-paradigm	LISP, Lua, Matlab, Python	• Easy integration with C and Fortran, scientific-oriented language • Several ML packages such as Mocha, or ML base

BIBLIOGRAPHY

Dole, V. S. Ethics in Marketing. *Vidyabharati International Interdisciplinary Research Journal*. 2019. (2): 155–157.

Feder, Judy. Machine-Learning-Based Early Warning System Stable Production. *Journal of Technology*. 2020. 72 (3): 59–60.

Federation of the Indian Petroleum Industry (FIPI). Annual Report 2021–2022. New Delhi. 2022. Chapter 5.

Medirata, Rajesh. "Accelerating the Pace towards a Market-based Gas Economy." Energy World. 2020 September 7. 4491. https://energy.economictimes.indiatimes.com/energy-speak/accelerating-the-pace-towards-market-based-gas-economy/4491.

Ministry of Petroleum and Natural Gas, Government of India. "Importance of Natural Gas for DPIIT Annual Report." Government of India, New Delhi. 2021. 11–12.

National Data Repository Report. Director General Hydrocarbon, Government of India, Ministry of Petroleum and Natural Gas Report. New Delhi, Government of India. 2017 July 28. Chapter 3.

NPCS Blog. *Handbook on Electric Vehicles Manufacturing*, Niir Project Consultancy Services, Delhi. 2022 June 15. Chapter 1.0–1.4.5.

Press Information Bureau Government of India, and Ministry of Petroleum & Natural Gas. "India Now Permits FDI in the Petroleum Sector across Hydrocarbon Industries." 2016 August 8. 148494. https://pib.gov.in/newsite/PrintRelease.aspx?relid.

Weatherl, Michael. Drilling Automation and Innovation, *Journal of Petroleum Technology*. 2021. 73: 57.

Index

Printed in the United States
by Baker & Taylor Publisher Services